NEIL
ARMSTRONG

NEIL ARMSTRONG

A LIFE OF FLIGHT

Jay Barbree

Foreword by

JOHN GLENN

THOMAS DUNNE BOOKS ✷ ST. MARTIN'S GRIFFIN NEW YORK

THOMAS DUNNE BOOKS.
An imprint of St. Martin's Press.

NEIL ARMSTRONG. Copyright © 2014 by Jay Barbree. Foreword copyright © 2014 by John Glenn. All rights reserved. Printed in the United States of America. For information, address St. Martin's Press, 175 Fifth Avenue, New York, N.Y. 10010.

www.thomasdunnebooks.com
www.stmartins.com

Designed by Steven Seighman

The Library of Congress has cataloged the hardcover edition as follows:

Barbree, Jay.
    Neil Armstrong : a life of flight / Jay Barbree.—First edition.
        p. cm.
    Includes Index.
    ISBN 978-1-250-04071-8 (hardcover)
    ISBN 978-1-4668-3634-1 (e-book)
    1. Armstrong, Neil, 1930–2012.   2. Astronauts—United States—Biography.
3. Project Apollo (U.S.)—History.   4. Space flight to the moon—History.   I. Title.
    TL789.85.A75B37 2014
    629.450092—dc23
    [B]

                                                                    2014008696

ISBN 978-1-250-04072-5 (trade paperback)

St. Martin's Griffin books may be purchased for educational, business, or promotional use. For information on bulk purchases, please contact the Macmillan Corporate and Premium Sales Department at 1-800-221-7945, extension 5442, or write to specialmarkets@macmillan.com.

First St. Martin's Griffin Edition: June 2015

10 9 8 7 6 5 4 3 2 1

For Jo

CONTENTS

ACKNOWLEDGMENTS

Principal Storyteller
Neil Armstrong

Editor in Chief
Peter Wolverton
Thomas Dunne Books; St Martin's Press

Associate Editor
Anne Brewer
Thomas Dunne Books; St. Martin's Press

Editorial Assistant
Mary Willems

Associate Publicist
Katie Bassel

Management
Martha Kaplan

Science Advisor
Princeton Physicist Dr. Gene H. McCall, Ph.D.
Chief Scientist Air Force Space Command (Retired)

Senior Advisor to the Chief of Staff U.S. Air Force (Retired)
Senior Scientist and Fellow, Los Alamos National Laboratory (Active)

Literary Arts Advisor
Bob Button

Spaceflight Advisor
Retired Chief Astronaut Charlie Precourt

Project Mercury
John Glenn
Alan Shepard
Deke Slayton

Project Gemini and Apollo
Neil Armstrong
Gene Cernan
Charlie Duke
Jim Lovell
Tom Stafford

Project Space Shuttle
Robert Crippin
Franklin Chang-Diaz
Brian Duffy
Robert "Hoot" Gibson
Rick Hauck
Charlie Precourt

Research
Nicole Gail Roberts

Pictorial
Bryce Barbree Harrington, Photo Editor
Photographic Restoration—Mark H. Widick, M.D.

Photographs
J. L. Pickering

Pictorial Research
Mike Gentry
Research Librarian, Johnson Space Center

Maggie Persinger
Research Librarian, Kennedy Space Center

With special thanks to contributors

Howard Benedict
Associated Press Aerospace Writer and Historian

Tom Brokaw
NBC News Space Anchor

David Brinkley
NBC News Apollo Anchor

Bob Button
Gemini and Apollo Spokesman

Martin Caidin
aerospace writer and space historian

Gene Cernan
last on the moon, *Apollo 17* commander

Colonel Bill Coleman
astronaut affairs, office of the secretary of the Air Force

Tom Costello
NBC News space correspondent

Walter Cronkite
CBS News anchor and space historian

David DeFelice
community and media relations, Glenn Research Center

Brigadier General Charlie Duke
Apollo 11 CapCom and *Apollo 16* moonwalker

John Glenn
first American in orbit and veteran U.S. Senator

Herb Gold
NBC space group's associate producer, Gemini and Apollo

Hugh Harris
director, NASA's public affairs and voice of Launch Control

Ed Harrison
chief of information, NASA's KSC

Jim Hartz
NBC News space anchor and *Today* show host

Lester Holt
NBC News space anchor and *Today* show host

James Holten
NBC News space producer and management

Chet Huntley
NBC News space anchor

Bob Jacobs
NASA deputy associate administrator, communications

Jim Kitchell
executive producer, NBC space coverage

Jack King
NASA *Apollo 11* commentator

Matt Lauer
NBC News space anchor and *Today* show host

Jim Lovell
Neil Armstrong's *Apollo 11* backup commander and
commander of *Apollo 13*

Scott MacLeod
Korean War F9F Panther carrier pilot and
Grumman lunar module instructor

Lisa Malone
director of media affairs, NASA's KSC

Ralph Morse
Life magazine's photographer of the astronauts

Frieda Morris
NBC News management of spaceflight

Danny Noa
NBC News space producer

Charlie Precourt
chief astronaut (retired)

Alan Shepard
first American in space and commander of *Apollo 14*

Deke Slayton
chief astronaut and director of flight crew operations

Lt. General Tom Stafford
commander of *Apollo 10* and Apollo/Soyuz

Russ Tornabene
NBC News space producer and management

Manny Virata
television networks press site coordinator

Bob Watkins
Grumman lunar module team

Harold Williams
NBC News Manned Spacecraft Center

Brian Williams
NBC News space anchor

FOREWORD

Neil Armstrong was an aviator. He flew everything heavier-than-air—even gliders. He never boasted, never lobbied to be the first to walk on the moon, yet it would be difficult to find another who achieved as much in a lifetime.

Once when we were talking about his first step on the lunar surface, I told Neil that I was not normally a jealous person, but for him I would make an exception.

He was never comfortable with fame, hated talking about himself, so how do you write a book about Neil Armstrong's life of flight?

It helps if you are Jay Barbree, a friend and a pilot.

Barbree covered every one of our spaceflights for the NBC networks, all 166, and Neil wrote, "Jay Barbree is one of the world's most experienced space journalists devoted to getting it right, and he does."

This is the story of Neil Armstrong from the time he flew combat missions in the Korean War, flew a rocket plane called the X-15 to the edge of space, saved his *Gemini 8* by flying the first emergency return from Earth orbit, and then flew *Apollo 11* to the moon's Sea of Tranquility. There he made the first footprints on someplace other than Earth.

If you turn the page, I think you will be pleased.

—John Glenn

INTRODUCTION

There was a time when Neil Armstrong was the most well-known person in the world. But despite the notoriety, he never used his fame for personal gain. As Brian Williams reported, Neil could have easily owned a chain of moon burger joints. He could have challenged the wealth of a Donald Trump. Instead he remained a private person until he died.

For someone who never considered himself special, everyone else did. The world praised him in all languages, every radio and television spoke his name, and newspapers and magazines everywhere heralded stories about his accomplishments.

In a quiet, critical way the fame annoyed Neil. He considered most of the stories were from the outside looking in. He very much wanted stories written from the inside, giving credit due others. He wanted not a book about himself, but a book about all flyers and astronauts—an interesting, moving, entertaining factual book about those who flew from decks pitching at sea, from runways hacked through jungles, from desert strips, and from snow-laden mountain slopes; those who rode rockets from launchpads to live on orbiting outposts; those who rode spacecraft to the moon and always found their way back. These were the people who punched holes in the sky, raced moonward, and Neil was convinced any one of them could have flown the same flights he did.

He regarded Tom Wolfe's *The Right Stuff* a great yarn and good filmmaking, but terrible history: the wrong people working on the wrong projects at the wrong times, a project that bore no resemblance whatever to what was actually going on. Neil said, "What else could you expect written by one who never got closer to spaceflight than a Manhattan penthouse?"

For a half-century, Neil Armstrong lived flight and space travel, and he and I talked about writing a book from the inside. But every time we got close Neil found he was not comfortable writing about himself. He could in no way brag his accomplishments were greater than others and finally, he told me, "Jay you write it. You're a pilot. You're one of us."

There are only a couple of things I truly know. First, as a pilot I couldn't carry Neil Armstrong's lunch box. Second, I was never going to find a way to drag Neil to a keyboard to write about Neil because he simply marched to his own drummer. He wasn't much for rubbing elbows. He was quiet without being shy with a well thought out fairness in his sense of humor and his brand of justice. By some people's measurements he was a recluse.

Over the years Neil and I did write together. Not about his notable deeds, but rather about most anything that did not praise Neil Armstrong.

He wrote the introduction to *Moon Shot*—the *New York Times* bestseller by Alan Shepard, Deke Slayton, and myself—and when he accepted invites before Congress to talk about the state of our space program he often asked for my input.

When Neil; Jim Lovell of *Apollo 13* fame; and Gene Cernan, last to walk on the moon, wrote an op-ed on what they felt should be the future of America's space program we reported their op-ed on Brian Williams's *NBC Nightly News*.

One of Neil's last projects was writing with me a five-story series called *Space in the 20Teens*. Five weeks after we finished the project, Neil Armstrong died at the age of 82.

In 2008, NBC News gave me a dinner for 50 years of service and Neil surprised everyone. The quiet man came.

Neil was the keynoter, and following the dinner we pulled up seats at the bar of one of our favorite watering holes from the past, and Neil confessed all the things NASA did to mislead the media during history's first walk on the moon. He and I closed the bar. I wrote the story and was instantly pleased it was read by more than two million.

Neil Armstrong's propensity for getting things right makes it essential we do our best to accurately chronicle this American icon's exciting life of flight.

I first met Neil in 1962, and I've researched every detail of his record-breaking years. I asked, and Neil and those who flew with him answered.

Left to right: Jo Barbree, Jay Barbree, and Neil Armstrong enjoy dinner with America's first in orbit, John Glenn, who is performing standup comedy out of the picture. (Courtesy of the author)

Neil's words in this book are direct quotes by me and others I know to be trustworthy. They are from years of personal e-mails, recorded interviews, 51 years of this reporter's notes and files—a voice history as Neil told two NASA historians and transcripts of every word he spoke in space and on the moon.

The primary engine driving this book is accuracy: that's how Neil lived his life. But I will not report to you a single word he told me in confidence. At times I will put myself in Neil's shoes to re-create his thoughts just as he conveyed them to me, and just as important, I shall write more in this book about Neil Armstrong than he would have liked.

NEIL
ARMSTRONG

Panthers launch for attack on North Korean targets. (U.S. Navy)

ONE

A WING AND A PRAYER

At daybreak, pilots came onto the deck of the aircraft carrier *Essex*. Bundled like schoolchildren on a snow day, they waddled to their jet fighters. One of the first to climb aboard was a young ensign named Neil Armstrong.

His plane's captain helped Neil into the cockpit—helped him connect his shoulder and lap straps, double-checked his parachute harness, and rechecked his oxygen mask. Last, he made sure Neil's life raft and radio were ready to go.

Suddenly, flight deck speakers blared, "Prepare to launch aircraft!"

The date was September 3, 1951.

Armstrong had turned 21 only four weeks earlier, but despite his age he was suited and ready to fly his seventh combat mission.

"Move jet into position for launching!"

The deck crew inched his Panther onto its catapult, and Neil reminded himself that his first few "hot catapult" launches had taken faith. There was a degree of uncertainty. If for one reason or another the catapult produced a weak shot, he and his Panther could end up in the water. He instinctively rechecked his harnesses and lap belt before resting a gloved hand on the jet's control stick. His other hand rested on the throttle. He was ready to increase his Panther's power plant to maximum thrust.

"Launch jet!"

Neil felt himself go rigid. He saw the launch officer whirl one finger above his head and he increased his jet's thrust to an unbearable roar and waited—waited until the launch officer whirled two fingers then moved instantly to maximum thrust.

The white heat scorched everything nearby and the launch officer's right hand shot downward. The catapult fired.

Neil felt his weight double. His facial skin stretched from the powerful acceleration. His lips lengthened. His eyes and body slammed against his seat.

Eight tons of jet and pilot were swept down the catapult track, and within one blink of his eyes—less than the length of a football field—Neil's Panther climbed from its carrier's deck. Acceleration held him prisoner as one jet fighter and its pilot reached for sky.

Minutes would pass before Neil Armstrong would fly high enough and fast enough to become part of the soul-lifting beauty of quiet flight. It was the perfect escape—streaking ahead so fast that the overwhelming whine of his jet never caught up. It was necessary to capture the moment—to enjoy the single last morsel of time wherein a fighter pilot could relax before sending his 500 pounders under his wings and his 20mm shells from his guns speeding toward their targets.

Neil was aware it was his final moment of sanctuary from those who would be trying to do him harm, and despite his age his older brethren of the air regarded him a competent flyer. He taught himself to focus on the reasonable and the plausible instead of imagined fears.

Neil would never lose sight of the fact he was a small town boy with 15 cents in his pocket. He was thankful he was flying jets for the Navy, thankful he had the get-up-and-go to study what made the machines he flew fly, and the good sense to hit the books until he roped himself acceptance into the Navy's Aviation Midshipman Program.

He had selected Purdue University. The Navy's seven-year program called for him to spend two years in the classroom studying aeronautical engineering. Then there would be flight training where he'd get his commission as a Navy ensign followed by active duty before completing classes for his degree.

But it didn't work out that way. After he had studied for a year-and-a-half the Navy discovered they were short on fighter pilots and Neil was called early to flight training.

He got his wings in August of 1950, two months after the Korean Conflict had begun. Neil was one of those rare birds, a midshipman with wings. He had to wait a few weeks for his ensign bar.

"I asked for the Pacific Fleet and was given the Pacific Fleet," he would later say. "I was first sent out to a squadron called FASRON, Fleet Air Service Squadron, which was a utility unit where I waited until there was an opening in Fighter Squadron 51 (VF-51). I'd be flying from the deck of the *Essex* with midshipman wages of 75 dollars a month plus flight pay calculated at 50 percent of my base wage."

Neil had no regrets as the morning flight continued, suddenly feeling himself enjoying the new day. The sun was peaking over the horizon, spreading its warming rays. He studied the sky filled with VF-51 pilots and their planes. It was a togetherness that brought with it a certain sense of safety.

Better than being alone Neil agreed with himself as he suddenly saw one of nature's stunning creations, Mount Fuji in the rising sun.

The magnificent volcano's cone was perfect rising 12,000 feet to poke a hole through the clouds. Just as suddenly as Japan's most recognized landmark appeared, Neil was aware the peace the great mountain brought with it was about to end.

Japan's Mount Fuji shows its peak above the clouds at sunrise. (U.S. Navy)

Dead ahead, across the Sea of Japan, were the mountains of Korea—ugly mountains placed there by some ancient geological event that tortured them into jagged boldness—left them twisted and scattered, presenting no organized or logical face to visitors.

Neil judged them as terrible mountains—mountains of pain and death that ran in crazy directions. Their peaks formed no patterns. Their valleys led nowhere yet somewhere. Hidden from his view were today's targets.

His group's assigned duty was to fly into a hot zone naval intelligence called "Green Six." It was the code name for a valley with gun sites, freight yards and trains, a dam, and one of those ever-loving stubborn bridges.

Neil was comfortable flying the F9F Panther. He thought of it as a very solid airplane—built by the Grumman team, the best airplane builders around. "But in retrospect," he said, "it didn't fly well. It didn't have particularly good handling qualities. Pretty good lateral directional control, but very stiff in pitch. Its performance both in max speed and climb were inferior to the Chinese MIG by a substantial amount.

"I'm sure I would not have enjoyed going against a MIG in my Panther," he laughed.

They crossed the Korean coastline. The guns were waiting. The *Essex*'s fighters began to descend in swift dips and dives to confuse the antiaircraft. Then John Carpenter, Neil's group leader, pounced his Panthers upon the heaviest guns with blazing fire, raking the big gun emplacements through grey smoke and bursts of flak as they thundered straight down "Green Six"—lower and lower they charged, releasing their 500-pound bombs as five-inch and three-inch guns, even machine guns, fired at their jets.

Neil was instantly aware that a single shell could pulverize any of them. When he climbed up from the valley, heaviness was upon his legs and his face

Panthers cross North Korea's coastline. (U.S. Navy)

was drawn down upon his chin. The gravity gods were at work as he kept his Panther snug on John Carpenter's wing.

Back upstairs Neil could see clearly the targets were essentially demolished with one exception—that damn bridge.

John Carpenter saw it, too, and their leader immediately recognized the job needed to be finished.

Carpenter rolled his Panther left and brought his group down again, jets screaming along the shimmering river. They were roaring toward the bridge like bats out of Hades, thundering straight for the stone and steel spanning the river. Neil quickly noticed it was historic—tall pillars rising above one of North Korea's major waterways and decidedly vulnerable. Neil activated his nose guns and watched his heavy bullets rip into concrete and stone before releasing his last 500-pound bomb, which exploded, tearing and twisting the bridge into useless steel.

Time to reach for sky again, Neil ordered only himself. He hauled back on the Panther's stick. Damn! He had only a brief instant to see an antiaircraft cable stretching hundreds of feet from mountain to mountain.

WHOMP!!!!!!

An ugly shock wave shook his fighter from nose to tail.

"If you're going fast, a cable will make a very good knife," Neil told me later, remembering how the tightly wrapped strands of steel had sliced through his right wing too swift to be seen. It cut metal, wiring, tubing, and his control connections. Instantly six to eight feet beginning at the wing's tip was no longer there.

Quickly Neil judged he was about 500 feet from ground; his speed was 350 knots. His damaged Panther was flying at an angle that could aerodynamically compensate for the loss of almost half of his right wing—as long as he held the undamaged aileron on his opposite wing at an extreme that it could compensate.

The aileron is the movable surface on an aircraft's wings that controls roll and the amount of bank needed to work in concert with the rudder to turn the aircraft. Neil had to make rapid judgments. Not only had he lost almost half of his right wing and much of its aileron, his elevators that controlled up and down pitch had become sluggish.

Damn! Neil spat. The ground was coming up, coming up fast, and he had to— *Oh, trim!* With lagging elevators, trim tabs would boost them so he could climb. As quickly as his thumb could move the "coolie hat" trim toggle atop his control stick he was rolling in trim to bring his jet's nose up. But nothing was happening! Wait, there it goes! His nose was rising. "Move, move," he shouted at the forward end of his Panther. And it did, just before he would have clobbered Korean dirt. He was headed back upstairs—a slow, steady climb—and he was instantly aware he was not breathing. He gulped in air. "Armstrong," he shouted aloud. "20 feet above ground is no place to be at 350 knots."

If Neil had had time to sweat he would have. Instead, the young fighter pilot radioed John Carpenter, "Hey, boss," he stammered. "I've lost . . . I've lost about half of my starboard wing. I'm carrying a lot of aileron to keep from rolling, and if I get too slow she's gonna' roll on me."

"Roger."

"I'm regaining altitude slowly," Neil told Carpenter. "I have all the trim back on its heels, and my elevators aren't much use. I'll have to make one hot landing."

"How hot?"

"About one hundred seventy."

"Too damn hot," Carpenter shot back, realizing the carrier couldn't handle 170.

"Yes, sir," Neil agreed. "But she won't fly any slower without rolling."

"Understand."

"Eject?"

"Eject," said the group head quietly, "and I'll stay with you all the way Armstrong."

The two flyers reflected on the decision they'd just made. Then Carpenter asked, "Think about 14,000 feet will do it?"

"Should," Neil agreed. "Just want to make sure I'm high enough to have time to complete all those ejection procedures before I hit ground."

"Good idea," Carpenter laughed, adding, "The nearest friendly territory is down south. It's Pohang Airport, K-3."

"The marines?" Neil questioned.

"That's the one."

"That'll be good," Neil agreed, adding, "But no bailout over North Korea."

"Yeah, not too many come back."

"If I miss K-3," Neil told him, "I've always liked water. It's a softer landing."

"Roger that," Carpenter agreed as the side-by-side jets climbed from Green Six located on a narrow valley road south of Majon-ni, west of Wonsan.

Neil wasn't alone in his thinking. Both he and John Carpenter were aware they were planning what most pilots viewed as one of the most dangerous parts of their job—ejecting at jet speeds. That was the bad news. The good news was Neil had confidence in Carpenter's judgment. He was an Air Force major, on an exchange with the Navy. He liked the challenges of flying off carriers, and Neil liked flying his wing—liked learning. But there was more—in his brief 21 years Neil had never really thought about bailing out or ejecting from a plane that could still fly. He had not trained to do such a thing. One of his classmates had gone over to parachute school in El Centro, California, for ejection training and had come back and told them how to do it.

That was the extent of Neil's schooling on ejecting and he began thinking through what he was about to do as the two Panthers continued south. The farther they flew the more the mountains of Korea showed beauty. Gone were the tortured profiles and the senseless chaos of the north. To their right reservoirs glistened like fine pearls holding the hills. To their left snow hung upon the ridgelines. Before them the waters of the Sea of Japan held their carrier. But the ship would not be there for Neil.

He would be ejecting so he studied the hazards. He had a shotgun-shell-powered seat to blow him quickly away from his plane. He would instantly be clear of everything around him, but that instant speed would slam his body with such force he would suddenly weigh 22 times his own weight, or in pilot lingo 22 Gs. Not a lot of tolerance for error! The likelihood of some kind of injury was high—injury to the shoulders, arms, legs, and feet if he was not properly tucked in. He'd best be in the correct position or the ejection could cause him to create a new crater in Korean soil.

John Glenn, a marine fighter pilot flying combat in the Korean theater, tells the story of his famous wingman, baseball slugger Ted Williams. Williams suffered an engine flameout in his F9F Panther and opted, despite that his jet engine might explode, to fly to the nearest alternate landing field. Ted Williams knew of a pilot who ejected and suffered permanent injuries to his feet. Ted Williams vowed to never eject. He feared it would end his baseball career. Famed test pilot Chuck Yeager, the pilot who first broke the sound barrier October 14, 1947, called ejecting from a speeding jet "committing suicide to avoid getting killed."

Neil smiled. He'd heard it all. But when you are left with one choice, you're happy to have it.

The science of the day on ejecting made it pretty clear that being shot out of a speeding jet could severely compress your vertebrae. This is why the Navy had put a lifetime limit on the total number of times one of its aviators could eject. Neil had no plans to test the limit. He reached for his ejection seat instructions and began reading carefully:

One: Reduce airspeed if possible. 250! Yep, that's about as slow as I want to go.

Two: Check that safety belts and harnesses are locked.

Three: Pull pre-ejection lever inboard and push hard down until locked. This jettisons the canopy, dumps cabin pressure, lowers seat, releases knee braces, and pulls safety pin in seat catapult firing mechanism.

Four: Pull your feet back and place them on their footrests.

Five: Sit erect, head back against the headrest with muscles tensed. Pull face curtain down until fully extended.

Six: After drogue chute opens and seat stabilizes, release the face curtain, cast off harness, and roll forward out of the seat. If altitude permits, delay at least five seconds before pulling the ripcord.

Neil took a deep breath and read a final warning on the instructions. DO NOT PULL RIPCORD WHILE IN SEAT.

Damn good idea, Neil thought as he moved his eyes from the ejection procedures for a last look outside. His squadron had all come back before with bullet holes in their Panthers. They'd patch them up, covered them with fresh paint, and they looked pretty good. This would be the first time he'd ejected, but with John Carpenter playing nurse on his wing Neil's confidence grew. He was ready. His calmness was inborn. During his adolescence, he'd had a recurring dream about flying. He would hover in the air and if he held his breath he would never fall.

Silently, across the sky the two jet fighters trekked south side by side. Neil was aware he and Carpenter had never been particularly friendly—their interests and ages varied. They never talked much, but now in the sky with sunlight gleaming on the mountains they seemed like best friends. In his earphones he heard Carpenter open his microphone.

"We'll make it, Armstrong," his lead said reassuringly.

"Yeah, no doubt," Neil quickly agreed, impatient to end any conversation that might prevent his thoughts from focusing on the task ahead.

Carpenter sensed Neil's unwillingness to break concentration. They moved ahead through the sunny spaces, nursing Neil's crippled Panther over North Korean villages. From time to time they could see a burst from a gun, and then suddenly they were there, moving over Pohang Airport with the sea in view. Carpenter told him, "Armstrong, make sure your shoulder straps and seat belts are tight."

"They're already choking me," Neil replied.

"Good boy," acknowledged Carpenter. "You ready to hit every item on your checklist?"

"Roger, I'm ready."

"You holding 250?"

"Roger."

"You'd better jettison that canopy right now."

"Good idea."

"See you back on ship," Carpenter said and Neil, with the loud snap of his canopy jettisoning and the blast of outside air whipping across his helmet, pushed as far back in his seat as he could. He firmly placed his feet on their footrests.

Muscles tight, he took a deep breath and shouted aloud his final checklist:

"Pre." (All was in place.)

"Pos." (He was in proper position.)

"Ox." (He'd switched on his small green bottle of oxygen that would keep him breathing on his way to the ground.)

Neil was blessed with the ability to put off fear until the predicament he was in was over. He reached up, grabbed the face curtain that would protect his upper body, shouted *"pull,"* and with one quick and firm jerk, he pulled the curtain over his helmet and face to the center of his chest.

WHAM!!!!!

He exploded!!!!!

Neil was blown out of his cockpit by a violent crack of thunder as he held the face curtain firmly over his helmet and eyes. The 22 Gs made him feel as if all his body parts had been squeezed into the space the size of a bread box and he felt himself tumbling head over heels backward.

He was aware his seat and his body were rocketing upward. Adrenaline pounded through every muscle. The small drogue parachute popped out and stabilized Neil and his seat in the airflow. He felt the blast of wind and the noise leave, and he was most aware his tailbone was hurting from the ejection's kick in the ass.

It was time to let the face curtain go. Now he could see sea and sky rapidly vanishing and quickly reappearing again, and he quickly caught his breath as he felt the Gs leaving his body . . . and he was suspended in midair . . . weightless . . . and he felt his seat become more stable. He literally ripped off his harness and rolled forward out of it—suddenly free with his ripcord in his grip. Despite his altitude he counted, "One Mississippi, two Mississippi, three

Neil's ejection from his F9F Panther was like the one shown here from this F9F Cougar. (U.S. Navy)

Mississippi, four Mississippi, five Mississippi," and pulled. The chute came streaming out gradually so as not to break his back. Then the best sight Neil had ever seen—a big Day-Glo orange-and-white main parachute blossoming above him. He felt safe, he felt well—he'd survived the high-speed ejection.

It was a great feeling to believe you were home free. Neil was happy to see his ride down was taking him back to land—back to Pohang Airport, K-3, where U.S. marines would have his back.

He floated earthward, swaying back and forth, and as he dropped lower, he could see a rice paddy below. He opened his helmet's faceplate and quickly disconnected all hoses and snaps. Then he removed his helmet, dropping it to Earth. Neil wanted nothing impeding his hasty exit from his chute just in case there were unfriendlies around.

He braced himself for—

Into the rice paddy he went with a good pop. Certainly not as bad as the kick he had just received ejecting from his Panther. He freed himself from the chute and began running for cover.

He had taken only a few steps when he saw his helmet and its straps. He stopped and picked them up, noticing his helmet was cracked from its fall.

He stood erect, surprised to see an American jeep racing toward him.

There was no doubt he had landed on the marines' K-3 base—driving the jeep was a face he knew. His smile was suddenly wider than half of Texas. It was one of his roommates from flight school, Goodell Warren.

Warren was now a marine lieutenant operating out of Pohang and he yelled, "Armstrong, what the hell are you doing in my rice paddy?"

"Goodie," he called smiling from ear to ear. "You never looked so good."

The two midshipman buddies grabbed each other and Goodell Warren told him the explosions they were hearing out at sea came from the mines in the bay the North Koreans were laying.

Neil suddenly realized if his parachute had not drifted back to land, he might now be afloat in those deadly waters.

But all was well. Lieutenant Warren took care of his friend Neil, taking him to the brass for immediate debriefing.

Neil would only spend one night with Goodell and the marines before being sent back to duty aboard the *Essex*. There he was greeted with some good-natured ribbing. One of his fellow pilots was John Moore, who would in years

With his parachute's ripcord still in his hand, a pilot tradition, Neil Armstrong tells his marine rescuers all about his ejection. (U.S. Marines)

to come be elected mayor of Cocoa Beach, Florida, the hometown of the launch site that would send Neil to the moon. Moore insisted that Armstrong pay for the navy property he'd destroyed including the helmet he'd cracked.

Neil had a good laugh, not in the least aware that the gods of providence were saving him for history.

TWO

TEST FLIGHT, HIGH DESERT, AND A SATELLITE OR THREE

The Lewis Flight Propulsion Laboratory in Cleveland in the 1950s was a much nicer place than North Korea's ugly mountains. It was perched on top of a lush green valley carved by the picturesque Rocky River. It stretched for miles to Lake Erie and was part of the thousands of acres of rustic parks that formed Cleveland's "Emerald Necklace."

That was on one side. On the other was, and still is today, Cleveland's Hopkins International Airport. The famed Lewis Laboratory is now the John Glenn Research Center, but before Neil could chase his dream of working and test-flying for such a research center, he had to transition from fighter pilot back to the classroom.

Most recognized Neil was more than a pilot with natural skills. He was smart, analytical, and blessed with an understanding of the technical. He returned to Purdue and cut himself a wicked schedule.

His alma mater in the 1950s was influential in aviation and he needed to complete his aeronautical engineering degree. It was no longer practical just to drive an airplane. You had to be much more, with an emphasis on research. There were the swept wings, Mach numbers, rockets . . . the lexicon of the budding new age, and Neil recognized it was not only exciting but a warning—exciting because of the promise of rocket and spaceflight that lay just over the horizon, and a warning because being a pilot wasn't enough. There was also this little thing about his earning a membership in the Navy's Aviation Midshipman Program. It cut the cost to pennies.

Neil doubled up on his classes. The outside world seemed to merge into a fog. He ignored it, logging one college credit after another.

Just as important, Purdue was well under way building the first university airport. Neil managed flight time to stay efficient, and once he had his studies well in hand, he was content to spend a day here and there lazily.

It wasn't that he hadn't anything grabbing his attention. There were plenty of coeds filling out tight dresses. One in particular was Janet Shearon. She had been a person apart from the others Neil had known and whom he'd dated. While neither of them ever had broached the subject directly, the assumption rested quietly that a future between them waited. Janet had never pushed, and even in her unconcealed deep pleasure at greeting Neil she was certain, as always, to keep their relationship unstilted and undemanding.

Neil had in fact told his roommate he planned to marry Janet. Of course given his tendency to keep his mouth shut, he never told her.

The summer before his last semester, Neil presented his resume to the National Advisory Committee for Aeronautics (NACA).

The agency was the forerunner of the National Aeronautics and Space Administration (NASA) and the world's premier group for aeronautical research. Neil not only wished to be a research test pilot and engineer at the NACA's Edwards High Speed Flight Station on California's high desert, he equally desired to play a major research role in the science of flight. Most assuredly, Edwards was where Neil Armstrong was convinced he fit.

There was just one problem. When Neil was ready Edwards wasn't. The High Speed Flight Station didn't have an opening, but what Neil didn't know, the cutting edge high-speed flight group didn't wish to lose him. NACA sent his application to its other centers. "The result was the Lewis Laboratory talked to me about filling an opening they had. It was the lowest paying job I was offered," he said. "But I think in retrospect, it was the right one."

At Lewis Neil was given an old and very slow airplane. The Air Force called it a C-47. Airline drivers called it a DC-3. The Navy called it an R4D.

An R4D was what NACA had, and the agency modified it as a flying laboratory. Neil grunted disapproval, but it was his—all his. It was perfect for studying winter ice buildup on wings and propellers. Neil soon learned the best way to study these icing conditions was during the winter months over Lake Erie. He flew his R4D into freezing rain and shed lots of the frigid stuff on the lake's icy waters. That was fun. Not the other part.

"The only product of NACA was research reports and papers," Neil grumbled. "Once research flights were completed, then you prepared something for publication. Next you had to face the 'inquisition.' This was a review of said paper by experts who were predominantly lady English teachers or librarians who were absolutely unbearably critical of the tiniest punctuation or grammatical error."

Good weather proved to be his savior from the inactivity of his flying lab. Clear skies lifted his spirits and lifted Neil into the air. He got to fly a faster, more agile F-82 in high-altitude flights over the Atlantic. In these tests he fired multistage rockets downward into thicker atmosphere for higher heat transfer rates.

Neil was just getting the hang of firing the multistage rockets when the call came from Edwards. NACA's High Speed Flight Station wanted him to come on board as a research test pilot.

They didn't have to ask twice. Research test pilot fit him like old clothes. Neil dropped his assignments at Lewis, and with his Edwards orders clinched tightly in his fist loaded his car and headed out. But there was one most important item. Janet. If he had his way she was to become his wife and the mother of his children.

En route to California he drove past Lake Erie and Lake Michigan, then headed north to see her at her summer job at a camp in Wisconsin. She was simply the most important person in his life, and the quiet one convinced himself he should invest a word or two in letting her know how he felt.

In the tradition of his true personality Neil thought long and carefully about what he would say. The words could be the most important he had ever spoken and he wanted to be absolutely certain he'd chosen the most sensitive and correct ones that would express warmth, caring, and meaning. "You know," he told Janet, "I've been thinking."

"Oh?"

"If you would marry me and come with me in the car to Edwards"—he was grinning widely—"NACA would reimburse me six cents per mile instead of four."

Janet laughed. "Sort of makes marriage worth considering, right?"

"Right," he laughed, "how about it?"

"Well . . ." She smiled as he watched the wind move around a few strands of her dark hair. "Perhaps it's a thought that should be considered."

Janet didn't ride with Neil to California that day, but she did marry him—a few months later in her hometown of Wilmette, Illinois, on January 28, 1956. As she had promised her parents her wedding had all the trimmings and proper protocols. Afterward, she and Neil drove their just-starting-out possessions across the Rockies to California's Mojave Desert, a barren, blistering place of sand and sun where Edwards Air Force Base sweltered.

The NACA's High Speed Flight Station hangar with its people and planes in the 1950s. (NACA)

Neil wasn't quite sure how Janet felt about the parched wasteland. He could only hope this place fifty miles east of nowhere would grow on her.

He was only sure he loved it. He didn't mind the heat. He didn't mind the desert and the bones of the foolhardy who had fallen where they met their demise. He had a few months to get used to living in Valhalla where the souls of the world's fastest pilots were received, where their longest runways were outlined on dry lake beds, where in the heart of this sprawling home of daring flyers huge aircraft hangars rose in the incessant heat and all creatures endured windstorms that huffed and puffed sand and desert waste across all it reached.

Temperatures crashed to near freezing at night, and in December the rains arrived. They laid down thin layers of water on the dry lake surfaces, oozing something prehistoric and wiggly to the top—ancient brine shrimp that sea gulls from the Pacific would dive in and gobble up. As the winds blew these thin layers of water back and forth, the lake beds below were smoothed and readied for baking by the hot sun. Then the great natural runways would return and the high-speed machines would fly higher and faster than ever.

Neil Armstrong believed Edwards was home. He settled in for what promised to be an exciting life having not a hint the place would be renamed on March 1, 2014, the Neil A. Armstrong Flight Research Center.

They were at the time building the X-15 rocket plane, but it would be four years before it would be ready to fly. While he waited Neil would stay busy as a lower-rung research test pilot and engineer.

"They wanted me to learn NACA techniques for collecting data," Neil said. "They had a P-51 Mustang, one of the fastest propeller-driven planes ever, and they made me go out there and fly it," he explained. "I had flown its double fuselage version the F-82 back at Lewis. It had very rudimentary instruments and data-collection techniques and I practiced lots of maneuvers for test purposes.

"It was good experience," he nodded assuredly.

For eighteen months Neil Armstrong flew every research aircraft NACA had—dropping the agency's manned experimental rockets, or chasing them in the fastest jets. NACA had them all—including two B-29s, the same aircraft that had dropped two atomic bombs on Japan.

"I flew both of them—right seat, or left seat," Neil said. "We used the B-29s to air drop over 100 rocket planes in the 1950s.

"The early drops were the X-1 and the Douglas Skyrockets, and I flew an average of ten times a month—just about every day we had good weather.

"Of course all the drops didn't go right—like the Skyrocket we dropped March 22, 1956.

"I was Stan Butchart's copilot with Jack McKay in the rocket plane to be dropped," Neil continued. "When we reached altitude in the B-29, about 31,000 feet, the number four engine quit, just quit running. Butch turned his seat around to help the engineer run down the problem and left the flying to me."

Neil took over the controls and they dropped Jack McKay in his Skyrocket. McKay landed safely on the dry lake bed below while Stan Butchart and the engineer busied themselves doing what they could do with the dead engine.

That's when it happened.

The B-29 shuddered. In that hair-raising moment Neil had a quick glimpse of a bullet-shaped propeller hub shooting backward. Engine number four's propeller had disintegrated. He was suddenly looking at a banged-up and dented engine without a propeller.

"It appeared the whole propeller shredded itself," Neil said. "And when it did, it clobbered engine number three."

Butchart and the engineer began trying to feather engine three's propeller—rotate its blades so they were parallel to the direction of the airflow. This reduced resistance and reduced the workload on the remaining two engines.

Neil was left with two good engines of the four, but not really. There was a problem. Both good engines were on the left wing, and they were approaching the north lake bed at 30,000 feet. Nobody did that—landing a B-29 from 30,000—and Neil was thinking get your gear down, create some drag, or you're going to overrun the lake bed. He was happy to see Butchart rejoin him at the controls.

That was the good news. The bad news was the disintegrating engine four had damaged their controls. Butch and Neil decided they could only use engine number two—just one of their four for landing; engine number one was the farthest out on the wing and had too much torque. It was already trying to wrench the controls from their hands. Somehow, with both Neil and Butch fighting their yokes and rudder foot pedals, doing their best to keep the plane from getting away from them, they managed a safe landing.

The B-29 emergency was no little thing, but Neil hardly spoke of it to Janet. To him it was another day of writing the textbooks on flight.

When it came to growing their knowledge, he and Janet were on the same page. They had found time to continue their college studies in Los Angeles area schools but there was another act of acquiring afoot. That next summer Janet gave Neil a gift. One he would care more about than flying. Their first child, Eric Alan Armstrong, arrived on June 30, 1957.

Friends immediately noticed a longer stride in Neil's walk.

He wouldn't brag on himself, but he wouldn't miss a chance to brag on the son he and Janet decided to call Ricky.

There was just one problem. Neil didn't have the patience to wait for Ricky to grow into a baseball glove. But he did manage—he waited and watched and his son grew, and Ricky eventually became the six-foot-something athlete his father wanted.

The strapping young Armstrong was a permanent fixture on the baseball diamonds and football fields. He soon lost the "y" at the end of his name—especially on the football team where they called him kicker.

In 2008, I introduced my grandson Bryce to Neil. I proudly told him Bryce had set football records kicking for the universities of East Carolina and Shenandoah. No sooner had the words left my lips the normally quiet Neil Armstrong was telling Bryce all about Rick's kicks.

Their conversation became lengthy—each unashamedly stretching the truth. I could only watch a talkative Neil Armstrong—yes, a talkative Neil Armstrong—brag on his first born.

What was a strong fact was that Neil loved Rick as he did all his children. There were things like family and flying he would go on about with those close to him—but not to the public. There Neil remained a mystery. His blue eyes seemed to reach all the way to his soul—they sent the message "keep your distance." His boyish face was absent of any lines, and his expression only changed with his distinct smile, his one-of-a-kind grin.

If you dared ask a question, Neil would stare at you, forcing you to wait and then begin asking again before he would finally speak. His words were perfectly thought-out sentences—a direct product of his scientific research training. Whatever he said he wanted it to be correct. He didn't like having to defend something his mouth had carelessly gotten wrong.

Most assuredly Neil Armstrong was big on fact and accuracy.

But it wasn't all work.

Once in a while there was fun. Neil would fly routine tests over his family's cabin and waggle his wings.

When he and Janet first set up housekeeping, they lived in a couple of places in the Edwards area before buying property in the San Gabriel Mountains.

"On that property on a mile-high perch was a cabin built for a weekend getaway, and compared to the comforts back home," Neil bragged, "the cabin qualified us for honorable mention in early pioneer folklore.

"Its floor was bare wood. No bedroom, just four bunks with a tiny bath and small kitchen with primitive plumbing and no electricity.

"Of course there was no hot water," he laughed. "The shower was a hose hanging over a tree limb. We did some remodeling. I ran some wiring to get us electricity, and Janet cooked on a hot plate.

"No one lived near Edwards," he explained. "It was a big base and you were always at least 30 minutes away."

The cabin was located where the flora was lush and the fauna were plentiful. "On a clear day," Neil assured us, "we could see across most of Southern California."

Northwest were the Tehachapi Mountains with their trails winding across the Mojave Desert floor to the fertile green fields of the San Joaquin Valley. To the northeast lay granite buttes amid sandscapes including Saddleback and Piute. And though the harsh summer sun baked the land below to a winter-wheat brown, in the wet springtime the entire valley bloomed into one of nature's perfect gardens.

"My job was 50 miles and one stop sign away," Neil laughed, not in the least aware that if you flew to the opposite side of the planet from his mountain home, you would arrive at another world not unlike Edwards.

You would be in the land of the sky: not the sands of the Mojave, but the vast steppes of Kazakhstan, a flat plain where the yellowed grasslands turn green only in the spring, where at days end one could see nothing, not even a leaf or twig between self and setting sun.

It was this bare, unpopulated land that was chosen in the 1950s by a small army of Russian space pioneers to build the Soviet Baikonur Cosmodrome. It was a sprawling space center located perfectly to launch rockets, where mishaps would do little damage to the sparse life surrounding it. Even more important, the desolation would hide the Soviet's secrets.

They developed and tested missiles and rockets just as Americans did at Cape Canaveral. And on October 4, 1957, the Soviets gathered around a large white booster for the first step on the road to arguably one of humankind's greatest achievements. They were planning the first of many major events needed to set in motion Neil Armstrong's exciting life in flight whereby he'd become the first to step on a place other than Earth.

The rocket was called R-7. Neither Neil nor anyone beyond its launchpad were aware that a historically momentous rocket was about to be fired. It would add greatly to the pages of history, and the man orchestrating the stop-and-go countdown was cut from the same bolt of cloth as those like Neil who asked the question, "How high is up?"

The man's name was Sergei Korolev and he was near the launchpad inside a steel-walled room. He sat at an old wooden desk, microphone in hand, directing his launch team. He was the chief rocket scientist for the Soviet Union, who, unlike America's genius in rocketry, Dr. Wernher von Braun, had the blessing and support of his country's government. His R-7 was four times more powerful than von Braun's Redstone, and it was about to send a satellite into orbit and open the road to the moon.

Korolev's simplicity would tolerate no fancy surroundings. Shortly after arriving in Kazakhstan, he built for himself a small wood-frame house not unlike Neil's California cabin. The essential difference between Armstrong's place in the mountains and Korolev's house was location: instead of 50 miles from Neil's runways it stood a mere walking distance from the R-7's launchpad.

Korolev left nothing to chance. He had worked side by side with mechanics and metalworkers, personally helping fashion and assemble what would be the first artificial satellite of Earth. Korolev created a sphere of aluminum alloys with four spring-loaded whip antennas and two battery-powered radio transmitters that would sing their unmelodious song to the world.

No science aboard this one. It was a satellite to simply demonstrate such a device could successfully be placed in Earth orbit, and he fitted it within a pointed metal nose cone and watched technicians installing it atop the R-7 booster.

Once the rocket's technological glitches had been resolved, events moved rapidly and they left the launchpad for safety behind thick concrete walls. The final countdown went quickly, heard only by the launch team, a handful of experts, and those officials protecting their place in the Soviet hierarchy.

An unsuspecting world was about to be shocked. The huge launch tower and its work stands were rolled back. The last power umbilicals between the tower and rocket separated, falling and writhing into their places of rest.

The rocket now stood alone.

The minutes were gone.

The final seconds were passing. Korolev's voice rang out: *"Zashiganiye! Tri, Dva, Odin."*

Enormous flame created a pillow of fire. It lashed and ripped into curving steel, followed concrete channels blowing long unbroken bright-orange flames across the desolate landscape. A continuing thundering roar followed. It rolled over Baikonur as Korolev's rocket climbed on an unbroken column of fire, delighting all that watched before leaving them and speeding away to reach for where nothing created by man had ever been. Korolev stayed inside. The Russian scientist was far more interested in the readouts from his rocket than seeing the startling, pyrotechnic display R-7 had created. He was not disappointed.

The numbers were perfect. Engines cut off on schedule. Stages separated as planned. Then, when the last engine died, protective metal flew away from the satellite. Springs pushed it free in space.

The satellite became known as Sputnik (fellow traveler). Obeying the laws of celestial mechanics, it immediately began to fall, beckoned invisibly toward the center of Earth. As fast as it fell in its wide, swooping arc, the surface of the planet below curved away beneath the falling satellite moving at a speed of 18,000 miles per hour in its orbit around Earth.

Some hour-and-a-half later it came back. Accounting for the movement of Earth beneath its orbital track in the time it took to circle the globe, Sputnik's path now took it fifteen hundred miles north of its still-steaming launchpad. It swept across Asia transmitting its incessant lusty *beep*. The loudspeakers of Baikonur blared its voice. Its launch team broke into cheers and shouts of joy. Korolev turned to them and spoke with deep feeling, "Today, the dreams of the best sons of mankind have come true. The assault on space has begun."

Neil Armstrong was in nearby Los Angeles at a symposium held by the Society of Experimental Test Pilots when Sputnik reached orbit.

He was disappointed Sputnik didn't belong to America, but found the Russian launch encouraging. "It changed the world," Neil said. "It absolutely changed our country's view of what was happening, the potential of space. I'm not sure how many people realized at that point just where this would lead.

"President Eisenhower was saying, 'What's the worry? It's just one small ball.' But I'm sure that was a facade behind which he had substantial concerns," Neil explained. "Because if they could put something into orbit, they could put a nuclear weapon on a target in the United States. The navigation requirements," he added, "were quite similar."

That said, in Neil's judgment the Soviet Union was without question technologically inferior. Someone had dropped the ball and it gnawed at him. He knew that Dr. Wernher von Braun and his Huntsville, Alabama, group were better. He knew von Braun's seasoned engineers had built rockets that were already reaching space, and America could have been in orbit with one of von Braun's Redstone rockets and a couple of upper stages long before now. Why weren't they? Secretary of Defense Charles E. Wilson.

Neil, as did many at the High Speed Flight Station knew Wernher and his group had been trying to get official approval to punch a satellite into Earth orbit for more than a year. But Wilson thought it was just so much nonsense and he and President Eisenhower were perfectly willing to put America's prestige on a larger version of the Navy's Viking RTV-N-12A sounding rocket. The Navy had tested about ten. About half had failed and the final launcher to carry a satellite was to have a larger engine and additional upper stages. In a sense the United States was putting its reputation on a yet-to-be built paper rocket named Vanguard.

Eisenhower and Wilson undressed it from its Navy whites and hung a sign around its neck that read "Civilian." It was now part of an international science project, the IGY (International Geophysical Year), which had a membership of sixty-seven nations. No one was sure the pencil-shaped thing would fly but it was the politically correct thing to do.

In 1956, Wilson ordered von Braun to remove Redstone 29 from its launchpad. It could have been a year ahead of the Russians in launching a satellite. But it wasn't. Neil and others on the front line of research wanted to go to Washington and ride Wilson out of town on a rail. Instead Neil went outside with the others from the test pilots' group to look for what was now orbiting Earth.

According to some, Sputnik could be seen sweeping over an early evening Los Angeles, but Neil knew better. There was too much reflected light in the metropolis's night sky for that. It might be possible to see the large trailing rocket stage glinting from the sun that was still lighting it above Earth's early darkness. The conditions would have to be just right. Neil and others looked for a while but, as he had expected, there simply was too much reflected glow. They gave up and went back inside.

Only a month after *Sputnik 1*, the Russians did it again. *Sputnik 2* raced more than a thousand miles above Earth. On board was a living, breathing animal. A dog named Laika.

Americans were livid. Was Eisenhower fiddling while Rome burned? Where were our rockets? Where were our satellites? What the hell was going on? The president got the message. He acted, but prematurely. A civilian launch team working on Vanguard rushed the unproven rocket to its launchpad. On top was a grapefruit-size satellite that was so small it weighed only a laughable three pounds.

The day was December 6, 1957. The launch team neared the end of its countdown and an anxious hush fell over a hopeful America.

"T-minus five, four, three, two, one, zero."

The slender Vanguard ignited, covered its pad with flaming thrust, and rose four feet, no more, before crumbling into its self-made fireball, consuming not only itself, but burning most of its launch facilities, leaving only blackened steel and ash.

The slender Vanguard ignited and rose four feet before consuming itself in its self-made fireball. (U.S. Navy and Air Force)

The loss of Vanguard wounded our pride, again. It also came close to destroying our confidence, and most Americans knew it was time for something to be done. The Russians were kicking us where we sat and it was time for a stubborn White House to call in the cavalry—to call in the von Braun team.

Eisenhower did, and Redstone 29 was hauled out of storage and refitted. A thirty-one-pound radiation-measuring satellite was mounted atop the rocket stack called Jupiter-C. The president and his White House didn't want to be reminded that the rocket was the same rocket that could have placed a satellite in orbit ahead of Sputnik. So the order came down to change the name and lessen their shame. The rocket would no longer be called Jupiter-C. It would now be called Juno-1.

On January 31, 1958, at 10:45 P.M. eastern the launch button was pushed. After waiting more than a year to fly, Redstone 29 came to life.

Yellow flame and thrust splashed outward in all directions. A huge pillow of dazzling fire gushed forth and thunder crashed across the Cape.

Those lucky enough to be there blinked at the searing flames and bathed in that marvelous roar. They cheered and screamed and some cried as the Juno-1 burned a fiery path into the night sky, reaching for von Braun's stars. One hundred and six minutes later, its satellite *Explorer 1* returned from the other side of Earth. America was in orbit.

A grateful and jubilant nation was at von Braun's feet.

Huntsville, Alabama, rocked with a wild and furious celebration. Horns blared and cheering thousands danced and hugged each other in the streets. Former defense secretary Charles E. Wilson, who had single-handedly stopped von Braun's efforts to reach Earth orbit, was hanged in effigy. Neil Armstrong was gratified. He was most happy von Braun's Huntsville group had proven America was and had been ready. He had a glimpse of the future. Perhaps pilots would not just be riding rocket planes across the skies. With the success of Sputnik and Explorer, pilots might soon be at the controls of spacecraft in orbit.

Cape Canaveral's sprawling rocket launch complex under a 1958 moon. (U.S. Air Force)

THREE

THOSE WHO WOULD
RIDE ROCKETS

Come the fall of 1958 Neil was surprised to see the new congressionally formed National Aeronautics and Space Administration's recruiters swarming about in search of astronauts for a new man-in-space project called Mercury.

One of Neil's assigned projects was the Dyna-Soar, a space plane he did not know then would become the forerunner of the space shuttle. It was a plausible idea, and to fly it, that project had earlier recruited astronauts. Nine were selected on June 25, 1958, for the Man-In-Space-Soonest (MISS) group.

"I was in the first lineup," Neil said, but with the formation of NASA, the Dyna-Soar astronauts were short-lived. The new space agency was starting all over and in October 1958 it set about recruiting the Mercury Seven astronauts.

Most who wished to apply hurried to Cape Canaveral, the place in those days considered vital, intensely exciting. It was in fact Florida's new dream attraction for tourists.

At night it was an all light show. Blinding searchlights surrounded its launchpads and blockhouses with their towering, shining rocket gantries. Support structures and hangars, even office buildings, were also awash with multicolored illuminations and soon it was obvious the bright lights were attracting the daredevils and the foolish.

But NASA rejected them outright, sending home the race-car drivers and mountain climbers along with all others from outside the pioneering family of aeronautics. The new space agency wanted the Neil Armstrongs, the John Glenns, the Alan Shepards—stable, college-educated test pilots screened for mental difficulties—not anyone willing to step outside of present-day accepted flight norms.

It was also unspoken that NASA did not want just experience. The agency did not want those getting on in years. This left out famed Air Force test pilot Chuck Yeager, who had had his day in the late 1940s and early 1950s.

Yeager broke the sound barrier October 14, 1947. What these new astronaut recruiters wanted more than a decade later was really NASA's own research test pilots like Neil and Scott Crossfield. These NASA pilots were flying all sorts of cutting-edge machines including the X-15, a rocket plane capable of reaching space. They were considered head and shoulders above their military counterparts by those who knew.

Unofficially Neil was asked to apply for Mercury. He found the invitation tempting, but he passed. He liked combining his engineering talent with test flights that had wings, and the X-15 had wings—not big wings, but wings, and even reporters were coming around calling it America's first spaceship.

The X-15 was really the most evil-looking beast ever put in the air. It was a 15,000-pound black horizontal rocket with little fins. It had a large blocky tail. Its black paint was there to absorb extreme heat generated by speed-induced friction in denser atmosphere. And best of all you could fly it—not into Earth orbit, yet, but that would come later with bigger rocket planes with heat shields. Neil simply could not warm to the idea of being strapped inside a capsule, a spacecraft like the proposed Mercury. It had no controls, no wings, no way to

Neil Armstrong and his stubbed-wing X-15 rocket plane. (NASA)

get you out of trouble bolted to the top of something trying to explode. But this was what NASA was building. A capsule you couldn't fly . . . but, oh, they *were* planning an escape tower with an instant rocket to snatch you away from a failing booster. Chuck Yeager had a name for those who would ride in it, "Spam in a can."

Top pilots from military ranks welcomed the NASA recruiters. After weeks of tests that froze, roasted, shook, and isolated them in chambers so quiet their own heartbeats boomed like the loudest drum in the parade, NASA selected seven. April 9, 1959, the agency introduced them in a news conference in the nation's capital.

They were called the Mercury Seven: Malcolm Scott Carpenter, a Navy lieutenant from the Korean War; Leroy Gordon Cooper Jr., an Air Force test pilot who flew the hottest jets at Edwards; John Herschel Glenn Jr., a Marine lieutenant colonel and fighter pilot from two wars; Virgil "Gus" Grissom, a flyer of 100 combat missions for the Air Force over Korea; Walter M. "Wally" Schirra, a Navy lieutenant commander, veteran of 90 fighter-bomber missions in Korea; carrier and test pilot Alan B. Shepard, a Navy lieutenant commander; and famed Edwards Air Force test pilot and veteran of 63 World War II combat missions over Europe and Japan, Donald K. "Deke" Slayton.

Four days after the Mercury Seven were announced the Armstrongs had an announcement of equal importance. Karen Anne became Janet and Neil's second child on April 13, 1959. They spoiled her, loved her, and Neil nicknamed her Muffie. Muffie brought a world of happiness into their mountain cabin.

For the next two years Karen grew into a happy toddler demanding much of big brother Ricky's time while their father Neil flew his X-15 higher and higher. The Mercury Seven astronauts hopped and jumped across the country, training and helping engineers develop and perfect the hardware they needed to reach orbit.

Cape Canaveral became America's early astronauts' favorite port. Its hard beaches were perfect for jogging, and if you were forever learning and training, where better to do it than in a warm winter sun, while those building your spaceship and rockets shivered in northern climes.

Falling in love with the Cape was not difficult. Even on the few occasions Neil's work brought him there he found the Florida spaceport's isolation equal to Edwards's.

There were the Cape's stand-alone complexes, thousands of electrical arteries, and a finely woven network of state-of-the-art computers, underground cable, and transmitters through which flashed the impulses and vital messages necessary for launching.

The astronauts loved it. They loved their beachside hideaway, and with the persistent mosquitoes, their smaller cousins the sand fleas, and other biting and crawling creatures under control, air-conditioning and tropical libations simply made the hot days and balmy nights a pilot's paradise.

But there was this continuing gnawing question: Who would fly first?

Then the day with the answer finally arrived.

The Mercury Seven waited at their desks. It was January 19, 1961. President-elect John F. Kennedy would be sworn in the next morning. But for the moment Robert Gilruth was more important to the Mercury Seven; he ran Project Mercury. "How about hanging in after quitting time, guys?" he called out to the men. "I have something to tell you."

John Glenn jogs on Cocoa Beach's hard sand beach. (NASA)

There it was. He'd decided, and the astronauts were grateful Gilruth got right to the point. "What I have to say must stay with you. You can't talk about it, not to anyone, not even to your wives."

He hesitated only to take a breath.

"Alan Shepard will make the first suborbital Redstone flight, Gus Grissom will follow Alan, and John Glenn will be the backup for both missions."

Six hearts sunk as the seventh raced ahead with pride.

John Glenn stepped forward and shook Shepard's hand. The other five moved in and offered their congratulations.

But before Alan Shepard would fly, there was a chimpanzee to launch. It was necessary to convince overly cautious officials that all was ready.

Ten thousand miles to the east on the steppes of Kazakhstan they did not take caution to its final step.

A man was trying to sleep. For a long time he had been drifting between slumber and wakefulness. This is how it had been for hours. He fidgeted with his thoughts, of what awaited him. He tried to escape by filling his mind with sights and sounds, with pleasing memories of his father—a carpenter, a skilled craftsman who had worked long hours to build their wooden home in the village of Klushino.

Those were the welcome memories, but he could not forget the nightmares— sights and sounds of a frightened boy covering his ears and eyes to block the memories of great guns blasting, shells exploding, the ground beneath his feet shaking from the rumble of the German tanks followed by even louder sounds. Airplanes. The high-pitch squeal of bombs falling toward them, exploding, taking lives, destroying homes.

At first it was only German planes. Then others came, planes with red stars on their wings—faster planes. The fighting was now up there, between the aircraft, and their fighting grew even louder while on the ground more tanks pushed into Klushino—Russian tanks. And finally when it seemed there was nothing left to kill, the war ended. But not the sadness, not the hunger—the memories of his parents foraging for food in the fields near their home, picking here and there to find a root or a missed morsel. He knew then the only way he could escape such a life would be with wings—wings like those with the Red Star.

Slowly the hunger and worthlessness were gone and peace and dignity returned. A young Yuri A. Gagarin was a permanent fixture in school and under the study lamp until he qualified to be a flight cadet.

In 1955 he entered flight training. Two years later he was wearing the wings of a jet fighter pilot. He became an expert parachutist as well, and after serving in operational squadrons and flying special flight tests, he volunteered in 1959 for an exciting new program.

Cosmonaut!

Four days earlier Yuri had been told, "You will be the first. You will be the first to fly in space."

It hadn't seemed real, but it was true. His backup Gherman Titov and the doctors came in, and everything went smoothly through breakfast before final medical checks. Sensors were attached to his flesh. He donned his pressure suit and heavy helmet. Gagarin was fully protected from the loss of atmosphere.

Titov helped him into the bright orange coveralls that would help helicopter recovery crews see him clearly when he parachuted onto the vast steppes.

The sun was up. It was bright. It bathed the huge rocket waiting to haul him into Earth orbit.

Yuri walked outside. It was April 12, 1961. He stopped and stood quietly for a long moment to study the enormous SS-6 booster. Then he waved and stepped into the elevator and rode it to the top of the gantry where his *Vostok 1* spacecraft awaited.

Technicians assisted him through the hatch, and once he was seated properly in his contoured couch they secured his harness and hoses and remaining connections. The cosmonaut waved a hand signaling he was ready. They closed the hatch, sealing Yuri Gagarin's destiny.

The countdown moved normally. Technicians took care of small problems that developed time and again, and then Yuri smiled. They had reached the final minutes that would end his waiting.

He relaxed his muscles when he felt motors spinning. The gantry was moving. Tall steel arms were pulling away and the launchpad was taking on the appearance of a huge metal flower. He felt the bumps and thuds of power cables ejecting themselves from their slots, leaving the SS-6 booster to draw power from its own internal systems.

The final seconds rushed away and Yuri heard the final word of the countdown: *"Zazhiganiye!"*

He needed no one to tell him his booster had ignited. He felt it. Twenty powerful main thrust chambers and a dozen vernier control engines ignited in an explosive fury of nine hundred thousand pounds of thrust—it shook him to his toes.

The mighty SS-6 rocket strained, explosive hold-down bolts fired, and the powerful booster was set free—the first human to head for Earth orbit was on his way.

It was mid-morning on the steppes of Kazakhstan. It was past midnight in a sleeping America. Only a select few at CIA listening posts heard Yuri Gagarin's jubilant cry, "Off we go."

Broad smiles grew into uncontrolled happiness inside the Baikonur launch control center as many whose duties were finished rushed outside to see the cosmonaut's rocket climbing into morning brightness. On board Yuri Gagarin was fully aware that he was now traveling faster than any pilot in history. Despite his weight being increased constantly by the pull of gravity, the first human headed into space maintained steady reports until he heard and felt a sudden loud thump, then a series of bumps and bangs as the protective shroud covering his spacecraft was hurled away. Now he could see clearly through his porthole. A brilliant horizon appeared, offering only blackness above; explosive bolts holding the central core of the SS-6's rockets fired, releasing the final stage to complete its job of hauling *Vostok 1* and Yuri into Earth orbit.

Suddenly with the SS-6's powerful first stage no longer firing, Yuri found he could now relax somewhat on his ride into orbit. He tried looking outside, but as he began trying to twist his body into position to see out the porthole, he heard the final stage fall silent, followed by more booms and thuds, and he knew *Vostok 1* was separating from its spent rocket.

Suddenly the miracle was at hand.

A human was traveling in space. He was moving through orbit at more than 17,000 miles per hour. As soon as computers on the ground caught up, he would learn that his orbit's perigee (its lowest point) was 112.4 miles soaring to a height of 203 miles at its apogee (its highest point).

Those on the ground listened in wonder at Yuri's matter-of-fact reports. He wasn't the least bit nervous, but they could not know that he was feeling like a stranger in his own body. He was not sitting up or lying down. In fact up or down no longer existed. He was suspended—no pressure from any direction on his body. He was floating, and only his harness strapping him to his contoured couch held him loosely in place. He was experiencing the magic of weightlessness—it appeared all about him in the form of papers, a pencil, his notebook, and other objects drifting, responding to the gentle tugs of air from his life-support system fans.

He caught himself drifting from his duties and he quickly turned his attention back to his schedule. He reported his instrument readings, but those on the ground were more interested in where he was.

"Tell us what it is you see comrade Gagarin?"

He smiled. "The sky is very, very dark. The Earth is bluish. You can see the atmosphere. It's like a blanket covering all the world," he told them, adding, "and all above it is black, dark black.

"The Earth below is mostly water," he continued, "but I can see some land. Most of it covered with clouds, but I can see no cities."

He thought to himself that from where he was he could see no evidence of humans ever having touched Earth. And then he raced into a sunset, into the darkest night he'd ever seen, with stars so abundant he could have never seen them all through the atmosphere. And he raced on, into the brightest sunrise he could remember, so bright he had to turn his eyes.

Suddenly he realized the passage of time. He was nearing the end of his one and only trip around Earth. He had to remind himself that he would only touch the controls if there were an emergency. He was more an observer than a pilot, but he would remain both physically relaxed and mentally vigilant as he monitored the automatic systems that were turning his *Vostok 1* around to fire its deorbit rockets.

Suddenly, he felt a kick that seemed to be sending him backward, that rammed him hard into his couch. But he wasn't going backward. It was the deorbit rockets blazing, slowing his speed, and he smiled with the full body blow. It meant everything was working. He had become a meteor plunging through the atmosphere across Africa. *Vostok 1*'s heat shield absorbed the high heat, and though he was inside a fireball he was cool and comfortable.

Then he was through reentry. The first spacecraft carrying a human had been slowed to subsonic speed, and at 23,000 feet *Vostok 1*'s escape hatch blew away. Yuri Gagarin was looking at blue sky, a flash of white clouds as small separation rockets sent him and his contoured couch flying away from the ship that had just carried him through an orbit of Earth.

Gagarin's parachuting skills came into play; his drogue chute billowed upward, everything was working perfectly.

For two miles he rode downward in his contoured couch; in the distance he could see the village of Smelovaka. Then, at 13,000 feet he ejected from his seat and deployed his parachute.

Yuri liked being a parachutist as well as a pilot and enjoyed breathing the fresh spring air, enjoyed the rest of the ride down.

On the ground a surprised man and his wife working in their field were watching. They were amazed to see Yuri Gagarin floating earthward in his bright orange suit and glistening white helmet.

The cosmonaut hit the ground running, tumbled, rolled over, and smoothly regained his feet to gather his parachute.

"Have you come from outer space?" asked the disbelieving wife.

"Yes, yes, would you believe it?"

Yuri Gagarin answered with a wide grin. "I have."

Neil Armstrong felt a definite kinship with Yuri Gagarin. He had no way of knowing how big the Russian cosmonaut's contribution would be to his future.

Yuri Gagarin is front-page news everywhere. (*The Huntsville Times*)

He only knew the world was jubilant over Yuri's accomplishment, while American newspapers crucified NASA for its lack of progress. Neil and all those who flew saluted Gagarin. Pilots were simply pleased Yuri was one of them. They knew the desire was born not only from watching great birds fly but from building models and finally your own wings.

Not far from where Orville and Wilbur Wright had built man's first successful airplane, Neil's father had bought him a ticket for a ride in an old Ford Trimotor. But like most pilots his real interest in flying began when he was old enough to ride his bicycle to his neighborhood airport.

It wasn't much of a bike. It had no fenders and he had to keep its tires pumped up, but it got him there—got him where he'd spend his spare time as a gofer for pilots who often rewarded him with a short, local flight. Some would even allow him to take the stick and fly short distances.

But that really wasn't the beginning. Flying for Neil began the day the man landed a gleaming new Luscombe at Wapakoneta.

There was no one around to help him service his plane so he called to Neil, "Hey, buddy could you give me a hand?"

Neil stumbled in his haste to get to the little ship. It was beautiful and Neil ran behind the right wing and pushed on the strut. The Luscombe was light and easy to roll along the grass to the gas pump. Neil helped the pilot fuel his Luscombe and the man said, "Thanks," and left for the flight line office to pay his bill.

When he returned he found Neil polishing the cockpit's side windows. The man stood silently watching for a moment.

Neil swirled his rag here and there, polishing the glass with care and the man asked, "Would you like a ride, son?"

Neil's grin answered, and once he was in the Luscombe's right seat he took a deep breath and soaked in the freshness. It was new. Everything was clean and sparkling. The instrument panel dazzled. Neil ran his fingers with care over the rubberized grip of the control stick on his side.

A few minutes later they were at 6,000 feet, drifting lazily above white puffs of clouds. Neil hadn't said a word, but his eyes were glued to every movement the pilot made. He couldn't believe it when the man turned to him and said, "Want to take it for a while?"

Once again Neil's grin answered.

The man laughed. "Remember, she's sensitive. Handle her gently."

Neil closed his fingers around the stick while the man held his hands up to signify passing the controls. Neil could not believe he was really flying the Luscombe. He soon realized he wasn't. He was manhandling it, and the man took back the stick.

"Hang on to it gently," the man instructed, adding, "Now, you follow my movements with your hand lightly on the stick. Let your feet rest on the rudder pedals." The man smiled at Neil. "Just follow through with me until you get the hang of it."

Neil nodded, and for the next few minutes he was feeling what flying was all about. He followed the gentleness of the man's touch—sensitive and respectful movements. He learned the Luscombe only needed to be nudged for a response. That's what the man was teaching him that day. Control the Luscombe with feeling, and for the first time, Neil felt like he could possibly be a pilot.

When the man was convinced Neil had the hang of it he asked, "You ever do any aerobatics, son?"

"No—no, sir, never did anything like that."

"Okay, hang on," and quickly the sky vanished. Neil was looking at the edge of the world. Where there'd been a horizon there were the green lands of Ohio, but only for a brief moment. The Luscombe was rolling around the inside of an invisible barrel until Neil realized the ground was up and sky was down. He tried to catch his breath when the nose went down and down, then up and up. An invisible hand pushed him gently into his seat as the engine protested, until the sun flashed in his eyes and Neil found himself on his back. The Luscombe had just soared up and over in a beautiful loop.

Neil's eyes were bright with delight and wonder and mostly knowing when the silver airplane whispered onto runway grass.

"That's when," Neil told me and Martin Caidin, "I really learned to fly."

He kept learning and even helped the mechanics doing engine overhauls, getting his pilot's license before he got his driver's license.

Neil filled his logbook by flying what was called slow time—repetitive slow flights around the airport to check out the repairs on aircraft that were long in years. Some dated back to the 1930s and had been patched together with faded fabric, their engines dripping oil. They smelled of gasoline in flight as well as on the ground, but Neil didn't care. He loved getting close to any aircraft.

He'd watch the pilots shouting "contact" to the mechanics, the plane's wooden propellers swinging down suddenly and catching with a stuttering cough. He loved standing behind the ships when pilots revved them up for power checks. The air blast whipped back, throwing up dust stinking of oil and gasoline. It flattened the grass, blew strong and heady into his face.

Neil loved the clanking, wheezing machines that required a certain care. They were old and they were worn and they created in him the need to know what made them fly. They were the driving force that forged his studies for an aeronautical engineering degree and the desire to be a research test pilot. When it was all said and done Neil understood and loved the fact that Yuri Gagarin was, too, one of his kind.

He was not in the least jealous. Envious? Why hell, yes! "Any pilot would be. Any flyer would have loved to have been first to orbit Earth," he said as he again questioned his decision not to file an astronaut application.

Deke Slayton filed his. Edwards never saw a better test pilot than Deke Slayton. He was a legend—at the top of his game—and he'd become a member of the Mercury Seven. And now with Gagarin's successful flight, Neil was left with little doubt the astronaut corps was worth reconsidering.

What Neil did not know, the morning following Yuri Gagarin's launch into Earth orbit, was that NASA was doing a lot of reconsidering, as well.

To quiet all the fuss over being beaten by the Russians again, NASA trotted the Mercury Seven out before the media. Saving face was obviously the plan. The astronauts showed their disappointment and offered sincere congratulations to the Russians for a terrific technical feat.

But once again it was honesty that saved the day.

John Glenn galloped to the rescue. He had a secret. Be blunt. Be truthful.

Neil smiled as he heard John tell reporters, "They just beat the pants off us. There's no kidding ourselves about that. But now that the space age has begun, there's going to be plenty of work for everybody."

John went over like rich cream, but others weren't buying it, especially a couple of second-guessers in the White House.

That morning John F. Kennedy was advised to forget about space. But the young president wasn't comfortable with that advice. How could he allow the United States to simply quit? America had come from behind before, and could do it again.

Kennedy's worst advice was coming from the head of his Science Advisory Committee, Jerome B. Wiesner from MIT. Wiesner wanted to gut the whole space program—cut NASA to its bare bones and reorganize from the ground up. Concentrate on aeronautics, he said, and yield the space race to the Russians.

Neil regarded Wiesner as the founder of the never-finish-anything-you-start crowd. Neil was comfortable leaving it in the hands of the president. He believed John Kennedy knew the American people better than Wiesner.

Neil proved to be correct. The president instantly rejected quitting. He knew his vice president Lyndon Johnson was a strong supporter of the space program. He called Johnson into his office and told him that from that moment on he would run the National Space Council.

Kennedy and Johnson agreed they needed a tough-willed North Carolina attorney named James Webb to take over NASA. Webb would be the agency's administrator. He knew how to navigate through government and industry and through executive and political pastures. Kennedy got right to the point. "Jim, I want you to run NASA."

Three weeks after Yuri Gagarin was first to ride a rocket into orbit, Alan Shepard climbed aboard his Mercury spacecraft *Freedom 7*. He was adding the fourth and final cornerstone to Neil Armstrong's structured destiny—a journey to the moon. Shepard was ready to become the first American and the second human in space.

Neil Armstrong was hanging onto every word coming from Mercury Control. None of NASA's research test pilots flying the X-15 rocket plane had come close to flying to the heights Alan Shepard and his *Freedom 7* spacecraft were about to reach. Neil was faced with the fact that if pilots were to fly through

Alan Shepard's Mercury-Redstone lifts off to put an American in space. (NASA)

President John Kennedy with first lady Jackie along with Vice President Lyndon Johnson followed Alan Shepard's televised launch into space. (The White House)

orbit and beyond as Yuri Gagarin had they would have to do so riding a space-craft launched by a rocket.

Neil clearly recognized Deke Slayton's voice as he sang out, "Five, four, three, two, one, *IGNITION!*"

"Roger, liftoff, and the clock has started," Shepard called back, instantly feeling the power. "This is Freedom Seven. Fuel is go. Oxygen is go. Cabin holding at 5.5 PSI."

Neil knew Shepard was in his element. He was as a test pilot the most relaxed, most assured person involved in the launch.

A rocket powerful enough to carry Shepard into space was climbing, racing skyward to new heights, rolling a deep pure sustained thunder across Florida's new spacecoast.

Neil was not the only one watching. There were in fact an estimated 45 million radio listeners and television viewers including some very interested spectators in the White House.

Aboard *Freedom 7* Alan Shepard was alone. He was reaching for sky fast and he was pleased. The flight was smoother than he'd expected. He radioed Mercury Control, "All systems are go."

America's first in space rocketed to a height of 116 miles and flew some 300 miles across the Atlantic before a huge parachute dropped his spacecraft into the sea and recovery forces plucked Alan Shepard and *Freedom 7* from the waves.

America still hadn't reached orbit as Yuri Gagarin had, but the U.S. was in space. Neil applauded. He had a decision to make.

FOUR

THE MOON IS CALLING

Alan Shepard's successful suborbital spaceflight had settled questions for President John Kennedy who accepted that Russian rockets and spacecraft were bigger. But he was coming to realize the Soviets weren't better because their technology could only build large nuclear warheads. They needed monstrous missiles to carry their monstrous bombs, but not America. With the significant breakthrough in size reduction in America's hydrogen bomb warheads, the same bang could be carried to any target by a rocket a third of the size. For this reason President Kennedy was convinced we were actually ahead of the Russians in rocketry, space vehicles, and the digital computer. He felt confident that in any technological race we could beat them. And Kennedy was ready to take what many considered a huge gamble.

Neil was in Seattle working on the Dyna-Soar project that day in May the president addressed Congress:

> I believe this nation should commit itself to achieving the goal, before the decade is out, of landing a man on the moon and returning him safely to earth. No single space project in this period will be more impressive to mankind, or more important for the long-range exploration of space, and none will be so difficult or expensive to accomplish.

There it was. Kennedy had thrown down the gauntlet of know-how and challenged the Russians. Congress leapt to its feet. Its members' applause shook the walls of the Capitol, and Neil Armstrong instantly felt the call. Kennedy wasn't just talking about astronauts orbiting Earth, he was talking about going somewhere humans had never been. He was talking about Columbus sailing to the New World, Lewis and Clark carving a trail to the Pacific Northwest, Byrd reaching for the North Pole. He was talking about exploration—stacking more wood on the stockpile of knowledge—and Neil instantly knew he would like to be a part of that challenge. But once again he was in the wrong place. Flying the X-15 to an altitude of 60 miles wouldn't get the job done. He had to become an astronaut. Quietly and instantaneously he moved to join the Mercury Seven as the launch team on the Redstone pad renewed its efforts to keep moving.

Gus Grissom's *Liberty Bell 7* Mercury spacecraft was ready to fly. On July 21, 1961, the second American lifted off and flew an almost exact duplication flight of the first. The splashdown was perfect, but that's where the duplication ended.

Gus was going through the drill of readying his capsule for recovery when an explosion blasted away his hatch.

Grissom's hatch had been modified to use an explosive primer cord instead of the mechanical locks on Alan Shepard's capsule. The primer cord inexplicably fired, and Gus saw waves coming into *Liberty Bell 7*. He scrambled out— swam for his life as frogmen tried to save his capsule. They failed but got Gus safely aboard the helicopter. Right away the experts began trying to determine the cause of the detonation. Some were sure the design of the capsule made an accidental explosion impossible. They insisted Grissom had to have hit the emergency plunger, which blew the hatch. "The hell I did," Gus snorted. "The damn thing just blew." The astronauts backed him all the way, and an accident review board cleared Gus of any wrongdoing. Four decades would pass before *Liberty Bell 7* would be brought up from the ocean floor.

Gus was right.

Sixteen days after Gus Grissom's flight Major Gherman S. Titov, who had backed up Yuri Gagarin, was sent into orbit. He stayed a full day.

Washington and NASA could only shake their heads.

It was obvious there was no longer a need to fly more Redstone suborbital flights. If America was going to reach orbit the same year as the Soviets and set foot on the moon by the end of the decade, they needed a bigger rocket to carry the Mercury spacecraft. It was time to bring Atlas to the launchpad.

The Atlas worked well boosting nuclear warheads 5,000 miles. But it was another story when it came to hauling astronauts. Atlas had no internal structure. Its strength was from inflation, much like a football. But its thin skin would collapse under the heavy burden of a Mercury spacecraft, much heavier than a nuclear warhead. It needed lots of fixing.

"Put a belt around its waist and it won't collapse," said famed rocket engineer John Yardley. Which they did: The steel belt held for chimpanzee Enos's orbital flight and they moved John Glenn's Mercury-Atlas to the launchpad while NASA went hunting for more astronauts.

With President Kennedy's challenge to reach the moon before the end of the decade the agency needed pilots to fly the Gemini and Apollo spacecraft. The Mercury Seven simply couldn't do the job alone.

The next group to be selected would be made up of nine astronauts. Like the Mercury Seven named for their spacecraft, the new group would be called the Gemini Nine.

Neil reached for an application, but as fate would have it, his small family was fighting a greater battle.

His infant daughter Karen Anne, who he'd doted over and nicknamed Muffie, was fighting an inoperable brain tumor. Neil and Janet had tried every specialist, had Karen Anne in every available medical facility, sought treatment and hopefully a solution from every corner of the medical world.

Neil's analytical and scientifically driven core would not permit him to believe there could not be a procedure to surgically remove the tumor. He searched everywhere, but as Christmas 1961 approached Karen Anne was becoming weaker. Neil and Janet got busy making their daughter's third Christmas special.

They did, and even though Karen Anne could no longer fully stand, she enjoyed her third holiday season.

Neil and Janet refused to abandon their search to make her well. Despite their unrelenting hunt for what would save her, their devoted efforts could not keep tragedy away from their door.

On Sunday morning January 28, 1962, Janet and Neil's sixth wedding anniversary, Karen Anne died. She succumbed to pneumonia and other complications brought on by the tumor.

The Armstrongs were devastated. Janet's emotions were uncontrollable. Neil's grief was a self-imposed quiet. Four-year-old brother Ricky was trying to understand. Friends and family gathered to comfort and help. They buried Karen Anne in the children's sanctuary at Joshua Memorial Park in Lancaster, California. A poem rested among the flowers: "God's garden has need of a little flower; it had grown for a time here below. But in tender love He took it above, in more favorable clime to grow."

The small stone marking her grave read: "Karen Anne Armstrong, 1959–1962." Between the two lines was carved, "Muffie."

NASA's High Speed Flight Research Center grounded all test flights the day Karen Anne was laid to rest.

Very seldom was Neil Armstrong not in control of his emotions. He would long for his daughter for years—no, for the rest of his life. He would never lose those special protective feelings he had for his little girl. Again and again he relived his inability to find the science, to develop it, to learn how he could have helped Karen Anne. In a large sense it came close to wrecking the man—a man who lived within the precise control of his abilities and limitations.

From the day Karen Anne was buried he could never pass Joshua Memorial Park without stopping, without visiting her grave. And yet in time Neil would come to accept the fact that science simply wasn't there when he needed it. No one on January 28, 1962, knew how to rid a body of an inoperable brain tumor. He didn't like it, but Neil reached a place where he could live with the fact that there wasn't anything more he could have done to save his little girl.

Eight years later during Neil and the *Apollo 11* crew's postflight visit to London, a two-year-old girl who came to see the spacemen was nearly crushed against a barrier by the adoring throng. Neil went to her rescue, then gave her a kiss. He clutched her safely until she could be returned to her mother.

The crowd of more than 300 cheered that moving moment. The next morning a London newspaper carried the headline, "2-Year-Old Girl Bussed by Moon Man."

Neil seldom spoke of this overwhelming heartache in his life. But others close to him were convinced Karen Anne's death was the single most important reason he would submit his name to become an astronaut. Her death gave him a new purpose. A few months before Neil's own passing I asked him, "Is there something of Muffie's on the moon?"

I read his smile to mean yes.

Only 23 days following his daughter's death the sun was coming into view at 6:47 A.M. at the Armstrongs' cabin in California. Three thousand miles to the east it was 9:47 A.M.—the sun was already shining brightly.

John Glenn sat atop his Mercury-Atlas ready to become the first American to rocket into orbit. Neil Armstrong sat before a television set. With the recent passing of his daughter he found it difficult to think about anything else. He had no way of knowing he was about to watch one of his future best friends soar into history.

John Glenn boards *Friendship 7*. (NASA)

"Godspeed, John Glenn!" the voice of astronaut Scott Carpenter boomed from the television. Neil leaned forward.

"Five, four, three, two, one, zero!"

Voices everywhere fell silent.

John Glenn's rocket was ablaze.

"Roger, the clock is operating," the marine reported to Mercury Control, "We're under way."

The sunlit Atlas-Mercury climbed from Cape Canaveral's famous rocket row as Neil focused on the spacecraft Glenn had named *Friendship 7*. It rested atop the flaming rocket and he could see the gimbals on the booster's main engines working in concert with the vernier rockets. He heard John say, "We're programming in roll okay."

Glenn quickly settled into his climb and just as quickly he and *Friendship 7* flew into Max-Q. Pressure squeezed his Atlas. The steel belt around his rocket's girth held. The marine fighter pilot reported, "It's a little bumpy along here."

He flew on into space. He was feeling what had been felt by Gagarin, Shepard, Grissom, and Titov. He now wallowed in weightlessness. He told Mercury Control, "Roger, zero G and I feel fine. Capsule is turning around. Oh"—Glenn shouted—"that view is tremendous!"

Glenn's Atlas-Mercury heads for orbit. (NASA)

America was in orbit and John Glenn settled in for three planned trips around Earth.

He knew the taxpayers who had sent him there wanted desperately to know what he was seeing.

Only minutes after reaching orbit he was witnessing his first sunset. He issued a glowing report. "The moment of twilight is simply beautiful," he told the millions listening. "The sky in space is very black with a thin band of blue along the horizon."

His eyes became acclimated to the universal darkness, and he turned down his cockpit's lights. He was now moving through the unbelievable black velvet, seeing so many firsts: a defined blanket of the brightest, most clearly defined residents of the universe; glorious stars, billions and billions of them; swirling galaxies, constellations, quasars, nebulae with their luminous, dark clouds and

John Glenn launches from rocket row. (NASA)

John Glenn reports, "The moment of twilight is simply beautiful." (NASA)

sprinkles of dust. And there were the planets, bold in the blackest of black skies. He could only stare in wonder at his first run through Earth's night side.

Glenn flew through the majority of his flight without a problem, but as he sailed through his third orbit consoles in Mercury Control lit up with a Segment 51 warning signal. It was telling flight controllers *Friendship 7*'s heat shield could have come loose. If so, extreme heat during reentry could cremate John Glenn.

Flight Director Chris Kraft and his team gave the warning priority attention. How could they save the astronaut? One idea quickly emerged. Survival might lie with the straps holding down the retro-rocket package.

The retropack contained six rockets. Three small ones had fired to separate *Friendship 7* from its spent Atlas rocket during orbital entry. Three larger rockets remained to decelerate Glenn's spacecraft, slow it so it would fall out of orbit.

The flight program was specific. The retros fired, the Mercury capsule slowed to start its reentry, then a signal was to be sent to break the metal straps. This would separate the retropack from the spacecraft's heat shield.

But what if you did not send the signal to break the metal straps? Would this not in turn hold the heat shield snuggly in place?

Flight controllers bought the plan. They had to do something to keep the first American to orbit Earth from returning as ashes.

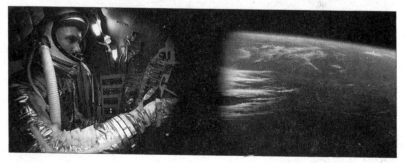

John Glenn could see Florida and Mercury Control dead ahead. (Composite photographs, NASA)

Once back over the Cape and Mercury Control, Alan Shepard was Glenn's capsule communicator and he gave the whole explanation to John for retaining the retropack. The marine understood the decision and told Alan Shepard to pass on his thanks.

"Roger, John," Shepard told him. "Hang tight, Marine. Navy has your back."

Friendship 7 and John Glenn raced around Earth on their final orbit, and when *Friendship 7* reached the California coast, the three retro-rockets fired. Glenn felt a triple thud and reported, "I feel like I'm going back to Hawaii."

Instantly, Glenn could sense the heat buildup. *Friendship 7* swayed. There was a bang behind him: part of the retropack breaking away. He called the Texas station. They couldn't hear him. He was plowing through an envelope of superhot ionized air. No signals could leave his spaceship. None could come in. All John Glenn could do was hold tight.

John Glenn rides through the life-threatening heat of reentry. (NASA)

America's first in orbit was cocooned inside a growing fireball. Glenn stared out his window at the flames devouring his ship. A strap from the retropack was burning freely, hammering against his porthole's glass. It burst into fire along with more flaming chunks that whirled away into space.

Then, he felt gravity forces building. He could have hugged them. That meant it was all holding together. He called Alan Shepard. He was feeling great, but there was no way to get through the ions. Not yet.

The heat shield on John Glenn's back was staying put. It was 4,000 degrees outside—toasty and comfortable inside. He now could smile.

In Mercury Control all listened intently as Alan Shepard continually called John Glenn. No response. He just couldn't get through. Notre Dame engineer Bob Harrington stood behind Alan Shepard, pleading, "Keep calling, Alan."

"*Friendship 7*, this is the Cape. How do you read? Over."

As instantly as they had come, the ions were gone and Shepard's call finally reached the Mercury capsule.

Glenn's reply was a simple mike check. "Loud and clear, Cape. How me?"

"Roger," Shepard acknowledged. "Reading you loud and clear. How're you doing?"

"Oh, pretty good," Glenn replied, "but that was a real fireball, boy!"

Mercury Control broke out in cheers and handshakes, and Harrington broke out with the Notre Dame fight song.

There was dancing in the aisles, but only for a moment. They had an astronaut and a spacecraft to land. *Friendship 7*, the little champ that it was, landed perfectly on waters near its recovery ship, *Noa*.

John Glenn gives President Kennedy a tour of his Cape launch site. (The White House)

John Glenn had returned a hero of Charles Lindbergh's stature. He had lassoed a share of the Russian lead, and President John Kennedy met him at his Florida launch site.

John Glenn receives a hero's welcome with a ticker-tape parade in New York City. (NASA)

When Glenn reached New York City, four million screaming, cheering people showered him, his wife Annie, and Vice President Lyndon Johnson with a tumultuous ovation, plus a hero's traditional ticker-tape parade.

Neil Armstrong leaned back in his chair. His low spirits from his daughter Karen Anne's death were lifted somewhat by Glenn's success. John, he thought, had taken another major stride needed for America to reach the moon. Neil was now certain that he wanted to be part of possibly history's greatest journey. He would, of course, continue to fly his X-15 assignments with his fingers crossed that he wasn't too late to join Glenn and the ranks of the astronauts.

John and Annie Glenn are honored with a parade in their hometown of New Concord, Ohio. (NASA)

Meanwhile the word went out from NASA. It's a long way to the moon. Keep the astronauts flying. Next in line Deke Slayton said, "Let's go."

But there were rumblings. There was a rumor about Deke's heart.

Presidential science advisor Jerome Wiesner, without a doubt the biggest hard-ass in the Kennedy administration, was at it again. The chief meddler spoke with the NASA boss. "Jim," he told Webb. "The White House has heard about Slayton's heart irregularity, and sending him into orbit could be a terrible mistake."

"How," Webb quickly asked? "He's been cleared by the flight surgeons and he's been flying . . . hell, he's been test-flying with this irregularity for years. What's the problem?"

"I know," Wiesner agreed. "But if something should go wrong, anything, and the word got out that Slayton had an erratic heart, who do you think they would blame?"

"The president," Webb agreed.

"That's right, Jim. Take Slayton off the flight."

Webb nodded.

Deke had idiopathic paroxysmal atrial fibrillation, a disturbance of the rhythm in the heart's muscle fibers in the upper chambers. The NASA chief called for a medical panel to review the facts. The panel agreed with Wiesner. The job of telling Deke was handed to his friend, flight surgeon Bill Douglas.

"Goddamn it, Bill, those sons-a-bitches can't do this to me," Deke shouted. "No one was concerned about this during selection. Hell, I've been flying the hottest jets out there—no big thing."

"I know," Douglas agreed, "But it's all about appearances. If something should go wrong, reporters would have JFK's ass."

"Instead it's my ass."

"Right! There's more bad news," the flight surgeon told him.

"What the hell now?"

"I know the rules call for the backup pilot to slip into the seat of an astronaut unable to make a mission," Douglas said, "but Wally won't be going."

"Why the hell not?"

"Bob Gilruth decided Scott Carpenter, Glenn's backup, has more time in the Mercury simulator than Schirra, so Carpenter will be going."

Deke turned away disgusted. NASA gave him a few minutes with reporters who had gathered. Deke did what was expected of him. He put the best face possible on possibly the worst news of his life. He took one for the team, then got the hell out of Dodge.

In the coming weeks Deke waged a fierce battle to return to flight status, and his fellow Mercury Seven astronauts rallied round him.

John Glenn stepped forward. "We're a team," he told the others. "Deke's still part of our team and we must give him his pride back."

"Yeah, man," Gus Grissom agreed.

"Let's make him chief astronaut," said Gordo Cooper, "but we'll have to hurry."

"Why's that?" Wally Schirra asked.

"The word is they're bringing in a general to take charge of us," Cooper told them.

"Like hell they are," Shepard, the future admiral, snapped.

"Let's take a vote and stand firm," Scott Carpenter suggested.

They all agreed that standing together would work, and they went to see the boss. "We have three recommendations for chief astronaut," John Glenn told

NASA administrator James Webb. "Deke Slayton, Deke Slayton, and Deke Slayton."

Webb smiled. The message was clear. He turned a thumbs-up, and Deke became chief astronaut.

Alan Shepard said it was like turning a switch. Deke's pride was back and first on his list was a new group of astronauts for the Gemini and Apollo projects.

Deke began reading applications and was pleased one was from Neil Armstrong. He smiled. He was going to have the horses he needed to ride to the moon.

Ready for launch, Neil Armstrong's X-15 hangs beneath its B-52 drop aircraft. (NASA)

FIVE

PASADENA OVERSHOOT

Dressed in his high-flight pressure suit Neil Armstrong was cocooned in his X-15's cockpit. The hatch had closed down on him to the point of being oppressive. His windshield wrapped his head and shoulders with two almond eyes that were set in a covering of Inconel X, a black-painted nickel alloy to dissipate heat. He felt as if he was wearing the cockpit instead of sitting in it. It was so snug it was difficult to see inside or out as he and his rocket plane hung beneath a drop-and-launch B-52. They were cruising at 45,000 feet—about 8.5 miles above the desert below. It was April 20, 1962, and as he approached his drop, Neil left the puffy white clouds behind, entering a CAVU (Ceilings and Visibility Unlimited) day over the dome of the world—what pilots called the high desert test-flight area.

The previous year Joe Walker, chief pilot for Neil's group, flew above 60 miles earning him the first set of X-15 astronaut wings. A reporter asked Walker, "How does it feel to be the best test pilot in the world?"

"Hey, I just flew a little higher than the rest," Joe Walker answered. "You looking for who might be best, keep your eye on young Neil Armstrong."

Neil had flown the X-15 six times. This was number seven, and like his seventh combat mission over Korea where he had to eject, this number seven would prove equally unlucky.

His flight plan called for him to take his X-15 to the edge of space, about 39 miles up, and test a new control system. They were now at that part of flight where things became tense—exciting—and if Neil Armstrong didn't know this feeling well, who in the hell did?

A clean and quick drop. (NASA)

The ten-seconds-to-launch light came on.

"Five, four, three, two, one, *launch*," and the B-52 dropped Neil's rocket plane, abruptly and with precision.

This X-15, the third in the fleet, had the newest and biggest rocket engine—the XLR-99, and Neil threw the switch. He felt it! The new engine's kick slammed him back in his seat.

At the family's mountain cabin Janet had the binoculars out. She could see the wide, sweeping contrails left by the B-52. She held her breath, wondering if Karen Anne's death would affect Neil's flying skills.

Janet was not the only one wondering.

Neil and his X-15 are on their way. (NASA)

The X-15 was not the easiest in the sky to fly. It was a nasty 51-foot-long black bullet with stubby wings. Its new, most powerful rocket was pushing it faster and faster from Earth. Neil's muscles tensed to handle the building G-forces. He was terribly busy, keeping everything under control while watching for the tiniest deviations.

He didn't have time to notice blue sky turn black, but when his big rocket shut down he knew he was moving—about 3,500 miles per hour, 1,500 miles per hour slower than the speed Alan Shepard had reached to rocket 112 miles into space. He would climb to less than 40 percent of that height. Now he had time to look out.

Neil's X-15 moves into black sky. (NASA)

The X-15 moved steadily upward on the energy it got from the XLR-99 rocket. This energy would push him through most of the atmosphere to where he could only see black sky above an extremely bright Earth below.

It was all breathtaking and dazzling and Neil could not believe the horizon's bands of color—colors that began with the deep blackness of space on top, then purple to deep indigo before settling into rich blues and bright whites, capping his planet's brown earth and blue waters. Neil was on his way still wondering how high is up as he left atmosphere behind and reached a place seen only by few, a place where air was so thin it could not reflect light.

The higher he climbed the less pull of gravity and sense of motion. He was now flying through the upper reaches of the sky—silent and weightless—an experience known only by those few who dared to sail over the top of the world.

"It was the highest I'd ever gone—thirty-nine miles and the views were spectacular," he later said, adding he was pleased with how well the X-15's new flight-control system was performing during his moment of weightlessness.

But soon the gravity grabbed him and his plane again, and started pulling

Neil's X-15 appeared to pass the sun on its way to a height of 39 miles. (NASA)

them back to Earth. Neil was controlling his X-15's attitude with tiny hydrogen peroxide jets near his rocket plane's nose. When needed, he would fire these jets to hold the X-15's attitude. Then, when he reentered enough atmosphere the X-15's stubby wings and flight surfaces would again take over the duties of attitude control.

It was all working well. So much so Neil momentarily diverted his attention to check the g limiter, a system he and other engineers had built to automatically prevent the pilot from exceeding five times his own weight. If it, too, worked, it could keep a flyer from blacking out.

Neil wasn't aware of how diverting his attention to check the g limiter was becoming a problem.

During the three months since his daughter Karen Anne's death Neil had seemed out of it to some. His ability to process facts had slowed. Some even believed Neil was becoming accident-prone, and in the flight-control center back at Edwards managers were watching Neil's performance closely. His primary attention for the moment was on checking the g limiter. He hadn't been watching his attitude closely, and as he descended through an altitude of about 27 miles, back through building atmospheric pressure, he noticed the X-15's nose had drifted up slightly. It was causing what was known as ballooning, a condition where the aircraft could skip off and along the top of the atmosphere like one skipping a flat rock across a lake.

In seven years, three months, and four days Neil Armstrong and his *Apollo 11* crew would be making the most risky penetration of Earth's atmosphere ever. They would be returning from the moon at 24,000 miles per hour—seven times faster than he had flown this day—and if their Apollo spacecraft did not hit the atmosphere at a precise angle they would skip off into eternity sailing forever across the universe.

Even though his momentary laxness had permitted the X-15's nose to come up, Neil wasn't all that worried. "In the process," he explained, "I got the nose up above the horizon. I was skipping outside the atmosphere again, ballooning, but that wasn't a particular problem." He used his reaction-control jets to roll over on his back and he tried a few other tricks but nothing worked. Mission managers in the flight control center were suddenly concerned, and the communicator with the call sign "NASA One" shouted, "Neil, we show you ballooning, not turning. Hard left turn, Neil! Hard left turn!"

"Of course I was trying to turn," Neil laughed, "but the aircraft was on a ballistic path. It was going to go where it was going to go."

There was only one thing to do—wait for the X-15 to fall low enough to get a bite of thicker air, and when Neil felt atmosphere, he'd begin his turn. But by this time, he said, "We had gone sailing merrily by the field."

There were those in flight control already calling Neil's mistake the biggest pilot error in the X-15 program. They were deeply worried. But not Neil. He knew he had options. He had altitude and he had airspeed and plenty of places to land. He concentrated on only one thing—get self and plane down safely. His years of flying had created in him an unbending drive toward perfection. That this was a state unattainable did not in the least interfere with his drive because Neil answered only to himself. He regarded excuses as a weakness and alibis as worthy only of disgust.

He was about 100,000 feet moving through the Mach 3 region (about 2,300 miles per hour), when suddenly he could see Pasadena. "Wait until Johnny Carson's 'little old lady from Pasadena' gets a look at this big black bullet?" he laughed aloud. "I bet she's never seen something in the air like us before?"

He checked out his location. Below he could see Lancaster and Palmdale, and the San Gabriel Mountains were straight ahead. His cabin was there. He questioned aloud, "What's next, the Rose Bowl?"

Neil rolled the X-15 into a bank and headed back toward Edwards where some in the flight-control center were already taking bets on where Neil would have to put the X-15 down. "How about Palmdale," one laughed. "They have a nice little runway there."

Later Neil would say, "It wasn't clear at the time I made the turn if I would be able to get back to Edwards, but it wasn't a big concern. There were other dry lake beds available." At the moment, El Mirage was off to his right, and just ahead on his left, was Rosamond dry lake. But, the prize of them all, the largest—Rogers—was dead ahead. It was home, and Neil was fast becoming convinced he could make it.

The boss, chief X-15 test pilot Joe Walker, was flying chase. Neil radioed him, "I have south lake in sight, Joe."

Joe Walker came back, "What's your altitude, Neil?"

"Got 47,000."

NASA One joined in. "Yes, we check that," confirmed flight control. "Have you decided what your landing runway is yet?"

Neil checked the ground beneath him. All he could see was desert and Saddleback Butte to his right. "Let me get up here a little closer. I can definitely see the base now," he told the flight center.

Suddenly Neil had his own cheering section. "He's going to make it," said one flight controller.

"I'll bet you a dollar he doesn't," said another.

"You're on."

"We show you 26 miles to the south lake and have you at 40,000," NASA One reported.

"Okay," Neil acknowledged, as he got busy ridding his X-15 of useless things that added weight and cut down his gliding distance. When he was done he keyed his mike and told flight control, "The landing will be on runway 35, south lake, a straight-in approach. I'm at 32,000, going to use some brakes to put her down."

Using speed brakes to reduce energy confirmed that Neil's return to south lake could not have been as close a call as some had thought, and the research test pilot, the calmest hand around, lined his X-15 up for a safe touchdown.

"I'm about 15 miles out from the end now," Neil told the center. "I'm 290 knots."

Neil opens his X-15's canopy with a smile. (NASA)

Fellow test pilot Henry Gordon was flying chase, too. He called Neil, "I'm coming up on your right."

"Okay," Neil acknowledged Henry. "I'm going to land in sort of the middle of the south lake bed," he told Gordon. "Brakes are in again, 280."

"Rog, start your flaps down now," Henry told him.

Neil's X-15's approach was as smooth as an Eagle gliding on mountain currents.

"Okay, you're well in," Henry told him. "Go ahead and put her down. The rocket glider made not a sound as it touched desert and Henry said, "Very nice, Neil."

And it was very nice. Neil Armstrong brought his X-15 in so smoothly its nose wheel barely kissed dry desert before rolling to a textbook stop.

Neil had just racked up the X-15's longest endurance mission (12 minutes, 28 seconds), and the longest distance flight (350 miles, ground track) a project record.

Could there be any more questions about Neil's ability to fly?

We think not.

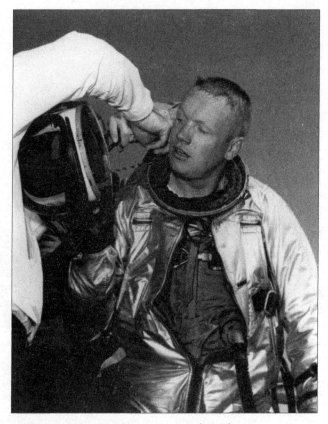

Neil Armstrong ready to be an astronaut. (NASA)

SIX

TRAINING DAYS

Project Mercury moved forward with the launch of Scott Carpenter. It was the project's second orbital flight, while back at Edwards Neil Armstrong waited. He was staring at the phone when it finally rang.

"Hi, Neil, this is Deke. Are you still interested in the astronaut group?"

"Yes, sir," he assured NASA's director of flight crew operations.

Deke was another man of few words. "Well, you have the job, then," he told Neil. "We're going to get started right away, so get down here by the sixteenth."

"Yes, sir, I'll be there."

Neil and eight others had been called by Deke. They would be announced as the Gemini Nine. Neil traded in his two cars for a used station wagon. He needed the larger vehicle to haul the family's personal belongings. He and Janet packed up the big stuff and furniture and shipped it ahead for storage in Houston.

There were still loose ends to be tied up so Janet stayed behind a couple of days.

Neil took Ricky with him in the station wagon, and two days later Janet took a commercial flight. She arrived in Houston the same day as Neil and Ricky and for the next few months the Armstrongs would live in a furnished apartment. They were waiting for their new home to be completed in the El Lago subdivision—a housing project built to attract astronauts. It was only a few minutes from Neil's new office.

The Gemini Nine were officially announced as NASA's second group of astronauts September 17, 1962. They were Neil Armstrong, Frank Borman, Charles "Pete" Conrad, James Lovell, James McDivitt, Elliot See, Tom Stafford, Ed White, and John Young. Some would become legends. Some would give their lives. None would be forgotten.

Neil thought of his group as answering the call for volunteers to fly to the moon. Their predecessors, the Mercury Seven, had been remarkable in converting "Spam in the can" to a successful human space program.

What NASA had in the Gemini Nine were test pilots, well-educated and experienced for the job. What Neil saw was *passion*, like most early NASA folks who were willing to work their tails off.

Deke Slayton now had fifteen astronauts under his wing. He set the new pilots up for indoctrination and training, and figured the more they saw of the remaining days of Project Mercury the better prepared they'd be for flying the heavier, larger, advanced Gemini. The Nine would be the ones to develop the flight maneuvers and procedures needed for that bridge to the moon.

Then, on October 3, 1962, the rookies gathered at the foot of the Mercury-Atlas launchpad. They watched Wally Schirra and his *Sigma 7* spacecraft thunder into orbit. The Nine hung onto every report from Schirra as the Mercury astronaut displayed his skills. The third American to orbit Earth stayed up for six orbits—nine hours—moving through his scientific and engineering check-list with an efficiency that would turn a robot green with envy.

The Gemini Nine were so impressed with Wally's flight they eagerly jumped into their grueling tour of Gemini's contractors. They spent weeks getting acquainted with where and how their Gemini spacecraft and Titan II rockets were being built and tested. Once they had an educated feel for the hardware they would fly, they settled in for their individual training program.

The good news was training for the Gemini Nine wouldn't be anything like that endured by the Mercury Seven.

They wouldn't have to drop their pants and sit on a block of ice and whistle while eating saltines. They wouldn't have to be stripped of all noises and sit in a soundproof room hearing only their hearts pounding.

"I don't think the community of flight medicine and flight physiology knew very much about what they needed at the time," Neil would tell me, adding, "There were widespread predictions that humans could not survive in space—for all kinds of reasons. I think they tested for everything, missing nothing," he grinned, asking, "Have you ever felt the need to hold a full enema in your colon for five minutes?"

We laughed and Neil had one more thing to say. "It was not fun."

Armstrong was content to spend his astronaut learning days with familiar exercises he'd experienced in naval flight training as long as he could study at night. "NASA felt we should know the difference between an aircraft and spacecraft," Neil explained, "especially what made you stay in orbit.

"I had studied orbital mechanics at the University of Southern California, and none of us found the new studies a burden."

They hit the books, and along with academics Neil and his group were called out for a number of other training disciplines.

To make sure they knew how everything worked, the Gemini Nine spent time crawling over and under Cape Canaveral's launch facilities in Florida. The bosses thought it would be good if they learned how rockets and spacecraft were launched, and how to get off of one of the damn things if there was a problem. They learned that, too, but first they had to learn how not to disturb the rattlesnakes and feral razorbacks that had made much of the launch complexes their home.

Then, when they had launchpad etiquette down, they returned to Mission

Control, Houston, to learn how they would be tracked and watched over in flight.

But no sooner than they were used to sleeping in their own beds again, they were off to Johnsville, Pennsylvania, to the Navy's Acceleration Laboratory to ride the "Wheel."

The Wheel was a huge centrifuge with a gondola, a mockup of the Gemini cockpit. This is where the astronauts experienced simulated reentries from Earth orbit. During these centrifuge runs, doctors would monitor them on closed-circuit television—one run simulated a G-pulling steep reentry. Neil was ready for this. None had anything close to his experience riding a centrifuge. When Neil felt the force of tremendous deceleration, he was able to keep moving his

The centrifuge wheel ridden by Armstrong. (Johnsville Centrifuge, NASA)

For jungle survival training in the Panamanian tropical rain forest, Neil and John learn how to build a dry floor for their survival hut. (NASA)

arms and legs until he approached 10 Gs (ten times his own weight). And as the g-forces mounted, his eyeballs flattened out of focus. When the centrifuge passed 15 Gs, Neil could no longer breathe.

The operators stopped the damn thing and Neil stepped from the gondola. He rechecked all body parts. He wanted to be sure he didn't leave any on the floor.

The Nine moved into additional training disciplines where they were taught how to wear pressure suits, and where they were introduced to weightlessness, vibration, and noise—even simulated lunar gravity.

John and Neil tough it out in their self-raised "Choco Hilton" in the Choco Indians' tropical rain forest in Panama. (NASA)

They also relearned how to use ejection seats and parachutes, and how to keep sharp their piloting skills and judgment by flying an assortment of jets that were based at nearby Ellington Air Force Base.

Then they were sent off with the Mercury Seven astronauts for jungle survival training to Albrook Air Force Base in the Panama Canal Zone.

They were dropped into the Panamanian tropical rain forest in pairs; Neil hit the jackpot. His partner was John Glenn and the two hurried hand-over-hand down a hanging rope from their helicopter into the jungle where they would build lean-tos out of natural materials and forage for food and water.

They were told the area where they'd been left was occupied by the Choco

Indians, natives still living as they had always lived. Neil immediately fashioned a welcome sign that read "Choco Hilton."

"It was anything but a Hilton," John told me. "We each had a knife and scant provisions, and we fished from a nearby stream.

"We learned immediately not to stick our hands under logs," he said with emphasis. "That's where the deadly *fer-de-lance* snakes live, along with lizards and other strange creatures we may need to cook over open flame—or eat raw."

Neil flashed one of his big grins. "For taste we topped our lizards and bugs off with a generous helping of worms."

Yuck!

Jungle training was the place where Neil and John became lifelong friends. Both told me it wasn't the best of places—so wet they could never find wood dry enough to make a really hot fire. They spent most of their time shivering and eating jungle delicacies cold, sometimes raw.

Each jokingly offered to trade the other for a woman astronaut. Both thought it would be great to snuggle in the wet cold with a bunkie. What they got were a couple of very ugly men—flight surgeon Bill Douglas and NASA photographer Ed Harrison. Neither came close to resembling a woman. They dropped by to make sure the duo was okay. Harrison took a picture to prove they were alive.

No sooner had Douglas and Harrison left when a painted Choco tribesman showed up with an assortment of trinkets and native doodads to trade.

The native admired their heavy writing pen, the one Neil used to create the welcome sign. He and John used gestures to talk the Choco local out of many of his trinkets before giving him the pen, then smiling and waving farewell.

They were pleased until Neil and John asked themselves: What if the pen runs out of ink before we're safely out of the area?

They sweated until it was time to walk back to civilization. They wisely followed the stream, cutting their way through jungle when necessary until they reached a river.

There they joined up with the other astronauts, and were ordered to put on their life vests. Some gung-ho survivalist thought they should spend a few hours in the water learning how to float safely down the river.

"What, no life rafts?" some questioned.

"You have one in your pocket," barked the survivalist instructor.

Neil and John kept quiet and followed orders. They donned those life vests, jumped in the river, and began floating toward their assigned pickup area.

"What about piranhas," yelled one concerned astronaut?

"The hell with the piranhas," yelled another, "what about crocs?"

Neil laughed and signaled John. "Hey, I think I see some piranhas over there," he said pointing, adding, "Steer clear of that area, John."

"Yeah," Glenn answered, "it looks like they're eating a crocodile."

Suddenly everyone in the river was quiet, looking, watching with trepidation the water around them. Neil and John smiled knowingly. Both knew the flesh-eating vicious piranhas weren't commonly found in the rivers of Panama, nor were crocodiles. Poisonous snakes like the large pit viper fer-de-lance, yes, and small alligator-like caimans, along with mosquitoes carrying malaria. But the chance of an astronaut being eaten alive was nil.

Their swim and float lasted three hours with all arriving at the river pickup point in one piece.

Neil logged the river float as one of the highlights of their training before they all parted in different directions. His future in space flight was ahead. John's was in the rearview mirror. President Kennedy, concerned an American hero of Glenn's stature might be injured or killed in another spaceflight, ordered him grounded.

Once back home Neil was suddenly concerned about another flight. The stork arrived April 8, 1963, with Janet and his second son. Of course the other astronauts wondered how Neil found time to get Janet pregnant.

The Armstrongs were all very happy with the arrival of Mark Stephen. Brother Ricky put aside his outgrown ball gloves and toys for the new baby.

Neil felt blessed. He and Janet had a healthy child. Neil was acutely aware that his first job was to make sure Janet and the boys would have everything they needed. Deke and the astronaut office stepped forward with that assurance, and Neil rejoined the Gemini Nine, where he and the group began riding parabolic trajectories in the undignified "Vomit Comet."

Neil said, "I left one source of vomit for another," and they all laughed possibly more at Neil making a joke than at the joke itself, and they went back to training.

The modified KC-135 was given the unflattering name of the "Vomit Comet" because vomiting was what its simulated weightlessness induced in some.

Neil could only smile. He had gone "over the top" in zooms in the F-104A

Starfighter at Edwards for the same weightless-inducing maneuvers without a problem. But he was surprised to learn those flights had failed to provoke the queasiness he'd feel with the Vomit Comet's abrupt changes in gravity.

Like most of the Gemini Nine, Neil endured four days in the Zero Gravity Indoctrination Program at Ohio's Wright Patterson Air Force Base. He and his new colleagues were introduced to floating free; tumbling, spinning, and soaring across the cabin by pushing off walls; eating and drinking in weightlessness, and using tools effectively in space.

Later, the new astronauts attended water safety and survival training at the Naval School of Pre-Flight in Pensacola, Florida. Naval aviators Armstrong, Conrad, Lovell, and Young had been there before. What was new for all Nine was learning how to stay afloat while awaiting helicopter rescue.

Neil found the transition from research test pilot to astronaut easy and comfortable. He recognized the similarities and appreciated the differences.

"We were looking for the best method we could find to go out and do something that's never been done," Neil said. "Being an astronaut is different yet the skills and the engineering and the hardware are really similar."

His astronaut brethren respected Armstrong's abilities as a pilot, engineer, and astronaut while most admired his intelligence. What set them to wondering was Neil's personality.

"Neil is quiet and thoughtful," Frank Borman said, "and when he says something, you think you should listen."

"Most of us came out of the same mold," added Alan Shepard. "But Neil is different. He's always trying to understand exactly what the inner workings of systems are—he wants to know what he's flying."

"Neil is as friendly as you can get," said John Glenn. "He's laid-back, a nice guy, small town just like where I came from. I don't think either of us put on any airs, and his sense of humor cracks me up.

"It's more British humor than mid-America," Glenn added. "Like the time someone told him, 'I passed by your house last night, Neil.' And Neil said, 'Thank you.'"

Dave Scott, who would fly with Armstrong on *Gemini 8*, had nothing but praise. "He's slow in making decisions on the ground but he's the quickest pilot I've ever seen in the cockpit. He's easy to work with, very smart—really cool under pressure."

The astronauts who would fly to the moon with Neil had something to say, too: "Neil is a very reserved individual," said Mike Collins. "He's more thoughtful than most, and in a world of thinkers and doers most test pilots tend to be doers. Neil is way over on the thinkers' side."

"Neil is certainly reserved, deep, and thoughtful," said Buzz Aldrin, who would walk on the moon with Armstrong. "He does not utter things that could be challenged later because of their spontaneity. In other words," Buzz laughed, "keep your mouth shut until you know what you're talking about."

The consensus among those who knew Neil was that he takes a long time coming to a decision on the ground and when he makes a decision that is that! But, say the astronauts, in the cockpit there's no pilot faster. He can read a problem and immediately correct it. That's Neil Armstrong.

Neil in the Gemini spacecraft learning the machine he would fly. (NASA)

———

Most believed the Gemini Nine's credentials were more impressive than those of the Mercury Seven. Neil Armstrong had not only been an astronaut in the first Man-In-Space-Soonest group, he had flown the X-15 rocket plane; Tom Stafford, Frank Borman, and Jim McDivitt had been instructors at the Air Force test pilot school; and John Young had aced two world speed-to-climb records—all of them were the best stick and rudder pilots around.

The short of it was that Gemini Nine belonged on the same field as the Mercury Seven, and on an upcoming trip to the Cape where they were getting ready to send Gordo Cooper into orbit for a record day-and-a-half final Mercury flight, the new astronauts came up with a great plan. They decided to show their respect and share fellowship with the established hands.

Henri Landwirth was the new space coast's beloved host.

Belgian-born, he spent much of his boyhood in one of Hitler's concentration camps before making his way to the United States with only two shirts, a pair of trousers, and one pair of shoes, a wardrobe that held up only long enough for him to land a kitchen job in a Miami hotel.

He washed the dishes, the pots, the pans, mopped the floors, cleaned tables, made beds, and smiled pleasantly at the guests. Then he became an American citizen.

Henri did every job there was in the motel and hotel business, and as a reward he was sent to Cocoa Beach. He ran his bosses new Starlite Inn, and when he learned the astronauts were not fond of their Spartan quarters on the Cape, for the charge of only one dollar per day, he became the astronauts' innkeeper.

Henri had a fondness, a protectiveness over those who flew through space, and he instantly recognized the need for one of his classic stunts to solidify the two groups brotherhood.

"You need to show your reverence and respect," he told Neil Armstrong and Tom Stafford. "Show the Mercury guys you're one of them. Bring them around to your side."

"How do we do that?" Neil questioned.

"The best way is to invite them to an evening of getting acquainted," he explained. "Host them for an unforgettable event. Serve them the finest wines and a cuisine befitting their station in life. Show them you are from the best families. You are cultured. You are refined. Show them your respect."

Refined my ass, Tom Stafford spat. His mother had to borrow the bus fare to send him to fulfill his scholarship at the Naval Academy so he could later become a three-star Air Force General. He had a far more important question: "Who in the hell is going to pay for all this culture and refinement?"

"Oh, the motel will," assured Henri. "We'll take care of everything."

Neil grinned quietly. He had come to know Stafford pretty well, and he knew the last time Tom picked up a Czech she was hitchhiking in Prague.

"Okay, then." Stafford nodded in agreement with the others.

"Let me handle it, guys," Henri told them. "It'll be the best of everything. Best food, best imported wines, the best."

Henri walked away smiling to set his plan in motion, and Neil looked at Tom. "I smell a dastardly deed here," he laughed.

"Me, too," chuckled Stafford. "Let's go help Henri make it more dastardly." And they did.

Under Neil Armstrong's and Tom Stafford's direct supervision, Henri printed a gold-leaf menu that called for a magnificent meal of breaded veal with au gratin potatoes, salads, and imported wines.

Each invitation was delivered personally by two of Cocoa Beach's sun-kissed beauties, who pointed out it would be black-tie (Henri rented tuxes, too).

The Mercury guys appeared the following evening dressed to the nines in their tux rentals. Showing their delight, they thanked Henri and their Gemini Nine hosts for what was promising to be a great evening of fellowship.

The imported wines lived up to their billing. They were poured, and words fell immediately to the required toasting and bestowing of best wishes and good fortune on one another.

Chief astronaut Deke Slayton heaped high praise on the Gemini Nine. Armstrong and Stafford returned the praise, making sure to show comradeship at its finest. It was the kind of togetherness that would warm any heart, and with the general high praise from each group quaffing the wine, 16 astronauts sat down to enjoy the gastronomical repast.

Waiters served the meal on Henri's best silver as silence descended with a crash.

Henri had prepared a sumptuous feast of breaded fried cardboard, uncooked potatoes, and a salad rotting from hours in the humid and hot Florida sun.

Suddenly, sobering silence gave way to hearty laughter that shook the motel's walls all the way to the swimming pool.

The Mercury Seven had been had—in the tradition of those who drive airplanes—with a classic "gotcha." It was just the ticket needed for the Gemini Nine to be warmly welcomed into the very exclusive astronaut club.

Gemini 5's prime and backup crews: Left to right, backup pilot Elliot See, backup commander Neil Armstrong, prime pilot Pete Conrad, and prime commander Gordon Cooper. (NASA)

SEVEN

HOME FIRE

S moke!

Why am I dreaming about smoke?

"Neil, wake up, Neil!"

Janet!

In that pleasurable place between sleep and wakefulness he could hear her. She was calling him and he could feel her hand on his shoulder. She was shaking him.

"Wake up, Neil, wake up!"

He rolled from his bed and set his feet on the floor. The smell was suddenly undeniable.

"Smoke, Neil, it's smoke!"

He leapt to his feet, bolted from the bedroom, but instantly recognized he couldn't see. His eyes were burning and there was heat and thick smoke and he wiped his eyes and swiped at the swirling burning fog. He managed to see a glow from the living room and with eyes throbbing and throat burning he yelled, "Janet, call the fire department. Call . . ." A spasm jerked his throat and he had to cough, and cough again, and finally he yelled a second time, "Call the fire department, the house is on fire."

Neil fought back the choking, the coughing, and quickly swirled about. First! What's first? Mark! Get the baby! He started to move but he couldn't breathe, he could only hold his breath—*hold it* he ordered himself. *Hold your damn breath*, and he began scrambling through the smoke and heat and the 3:45 A.M. darkness and then he was in Mark's room. He rolled the baby in his blanket and secured him in his arms. Fighting the smoke's burning acidity, he managed to bring Mark back to their bedroom but Janet was gone. He could hear her outside, calling their next-door neighbors Pat and Ed White.

Ed White, the West Point athlete and astronaut who in little more than a year would become the first American to walk in space, was quickly over the five-foot fence between their yards ready to help.

Neil was standing there, out of the house with his arms outreached, handing Mark to Ed. "Take the baby," Neil told him, spinning around and quickly grabbing a towel. He wet it and disappeared into their burning home again.

"Ricky," he yelled as he wrapped the towel around his face and dropped to all fours. The inferno was growing and he began scrambling forward through the roiling smoke and under the flames that were curling down from the ceiling.

"Ricky, where are you, son?" he called, feeling his way move by move to his eldest son's room. If only Ricky would scream or make some noise it would be easier, he thought. Then, suddenly he feared the worse. What if Ricky couldn't scream? What if he can't yell? Suddenly Neil was like a terrified spider moving beneath the smoke, scrambling past the flames until he was finally by Ricky's bed.

He was okay. Ricky was okay.

He grabbed his terrified six-year-old and wrapped the towel around his face, and beat a record-setting retreat below the flames and through the heat and smoke until they were safely outside.

Janet grabbed Ricky. Ed White was busy fighting the fire with a garden hose. He had passed Mark over the fence to his wife Pat who had gotten through to the fire department, and now the living room wall was glowing red. He could hear the cracking of window glass mixed with the sound of the roaring flames, flames that were causing wood to explode with pistol-like shots telling Neil the fire was devouring their home.

The driveway was growing hot beneath his feet and he heard sirens coming, saw the fire trucks flashing lights. Hooded men with masks ran straight for the burning house, and Neil yelled, "I got them. I got them all. There's no one inside."

The fire department took over with Neil and Janet and Ed and Pat following the firefighters every move as the professionals fought the flames for more than two hours. That's when night left and it appeared the last ember had been drowned. The sun came up and everyone could see the mess.

The Armstrongs, the Whites, and the other neighbors got busy. Even some fire-fighters stayed to help rescue what could be salvaged, and together they moved everything into the Whites' yard and under their carport.

Ed and Pat White took the Armstrongs in for a few days and neighbors and people Neil didn't even know brought over toys for the boys. One had a playpen for Mark and another a crib for him to sleep in, while others stacked up diapers and things they said they didn't need and Neil thought it was a tear he had just wiped from his cheek. The neighbors and other concerned people had brought all kinds of things the Armstrongs might need until they could move what had been worth saving into a nearby rental house.

It wasn't lost on Neil that the fire could have been a catastrophe for his family if they had become asphyxiated before Janet woke up.

The scientist in him wouldn't let him forget. Neil had to know what caused the fire. He had to know how to keep it from happening again.

During his urgent trips into the burning house to get Mark and Ricky, Neil had a sense the fire had started in their large living room with its high cathedral ceilings and beams.

The builder had used standard drywall-frame construction and the Arm-strongs had the builder put paneling over the drywall.

Within weeks the panel began warping, curling up because of moisture, and the paneling no longer fit. Neil had the builder back, who admitted to his mis-take, and his carpenters used a nail set to pound the finishing nails farther into the drywall and studs. The warped paneling simply fell off so the builder then put up sealed paneling and the Armstrongs thought their problem had been solved.

After the fire, inspectors found the cause instantly; one of the nails holding the warped paneling had been driven so deep into the wall it cut through the insulation of an electrical wire creating a trickle short. The small flow of electrical current built up the temperature in that location and when there was flammable heat, materials within the wall ignited. Unfortunately, at 3:45 in the morning.

The Armstrongs selected a different builder, a fire specialist, who built from the roof down instead of from the ground up. In the neighborhood, fire detectors were installed beginning with the Armstrongs and then neighbors Pat and Ed White, Faith and Ted Freeman, and Marilyn and Elliot See.

Within three years of the Armstrongs' close call with their home, four astronauts from the neighborhood would be killed. Ted Freeman's T-38 would fail him when it crashed at nearby Ellington Air Force Base October 31, 1964, and the prime crew for *Gemini 9*, Elliot See and Charles Bassett, were killed when their T-38 struck the roof of the building housing their spacecraft at the McDonnell plant in Saint Louis. Astronaut See was also the pilot for Neil Armstrong's backup *Gemini 5* crew, and the Armstrongs' good friend and next-door neighbor astronaut Ed White would give his life in the *Apollo 1* launchpad fire January 27, 1967.

It took the builders most of 1964 to rebuild the Armstrongs' house. They moved in a couple of days before Christmas, but Santa didn't bring anything resembling normalcy. Only time could bring them that.

The El Lago astronaut neighborhood began when the Whites and the Armstrongs arrived in the Houston area as members of the second group of astronauts. Ed and Neil bought three lots together in the development—one of several planned communities to house those working at the new and thriving space center.

El Lago was principally a neat, ranch house community. It had crisscrossing streets, and the Whites and Armstrongs split their three contiguous lots so each would have a lot and a half.

Other members of the Gemini Nine, the Staffords, Bormans, and Youngs built homes in the El Lago subdivision, too, while the Mercury Seven's Glenns, Carpenters, Grissoms, and Schirras lived at nearby Timber Cove.

The two neighborhoods were virtually an astronaut colony that would be strengthened by new members from the third group of astronauts selected in October 1963.

Director of Flight Crew Operations Deke Slayton now had thirty astronauts under his wing.

Eight of the fourteen new space travelers were test pilots. Five of those test pilots were from the Air Force. They were Donn F. Eisele, Charles A. Bassett, Michael Collins, Theodore C. Freeman, and David R. Scott. There were two test pilots from the Navy. They were Alan L. Bean and Richard F. Gordon, and one remaining test pilot was from the Marine Corps. He was Clifton C. Williams.

The remaining six were all pilots, but more important they came with wide-ranging backgrounds from academia. Edwin E. "Buzz" Aldrin Jr. had a doctorate in astronautics from MIT. His dissertation on orbital rendezvous would be essential in the flight sequences needed to reach the moon, and his dissertation would surely come in handy when he and Neil Armstrong walked on the lunar surface July 20, 1969.

Joining Buzz from the halls of learning was Air Force fighter pilot William A. Anders who held a master's in nuclear engineering. Another newcomer, Navy aviator Eugene A. Cernan, had an engineering degree from Purdue and a master's in electrical engineering from the U.S. Navy Postgraduate School. In fact, Cernan would be the last astronaut to walk on the moon. Former naval aviator Roger B. Chaffee also had an engineering degree from Purdue. Then there was Walter Cunningham, a former marine fighter pilot with a master's degree in physics from UCLA, and former Air Force pilot, Russell L. Schweickart, with a master's in aeronautics and astronautics from MIT.

Neil didn't know it at the time, but this third group of fourteen astronauts had just brought him Dave Scott, who would be his crewmate for *Gemini 8*, and Buzz Aldrin and Michael Collins, who would fly with him on *Apollo 11*.

Late 1964 was also the time this writer really got to know Neil.

As a member of the Cape Canaveral press corps I had been covering him and the other Gemini astronauts for two years. But beyond a brief hello and a question or two, our association was scant at best until mutual tragedy was the catalyst for a friendship that would last 50 years.

My wife, Jo, and I had a son born five weeks premature on November 22, 1964. The local hospital failed to take proper precautions. Our baby developed Hyaline Membrane Disease and we did everything we could to help the little fellow develop his underdeveloped lungs.

I had to keep an appointment with longtime friend John Rivard, and during our conversation I had a mental image of Jo sitting up in her hospital bed, crying. She needed me. "I have to go," I said, interrupting John. "The boy just died."

"I be danged," John said, adding as I walked away to my car, "If I can help, let me know."

There were no cell phones in those days—only landlines and long waits on switchboards and such. I drove as quickly as I could to the hospital, and then ran down the hall into Jo's room.

She was just as I had seen her in my mind, sitting up in her bed, her face awash with tears.

"Scott's dead, Jay," she cried.

"I know," I said, "I got your message about ten minutes ago."

We comforted each other, and Jo remained in the hospital a couple of more days. Our three-year-old daughter Alicia was with Jo's parents in Orlando.

The next morning I was having breakfast at a local restaurant when Neil Armstrong walked in. I mumbled a hello and motioned him to sit down.

He noticed right away that I was down. "Did someone shoot your dog?" he asked.

"It's a bad time for me," I explained, telling him about the death of our son.

Neil was from a small town of 6,000. I was from one even smaller, 5,000. A child's death was a tragedy generally shared throughout those small communities.

Neil told me about losing his daughter Karen Anne, and despite my tragedy I immediately realized that having a child for more than two years, and then losing her, had to be even a heavier burden.

We both considered ourselves lay members of science, and after a while we were trying to analyze the message I had received that my son was dead. Was it mental telepathy from Jo? Was it imagination? Was it only a coincidence? Was it heaven-sent? Whatever it was, Neil argued, "There is much that cannot be explained today. Not by science. Not by any sound reasoning," he said as a matter of fact. "Be thankful you had such an experience."

From that day forward Neil trusted me with information I believe he would not trust with others. It was understood I would not report anything without his permission. I have never knowingly broken that confidence.

In February 1965, Neil received his first assignment to a flight. He had been doing what he was told and had never lobbied for a mission when Deke Slayton named him Gordon Cooper's backup commander for *Gemini 5*.

The quiet one was quietly pleased.

Gemini 5's primary mission was to stay in orbit eight days to prove astronauts would have no problems living in space long enough to fly to the moon and back. "Eight days or bust" was *Gemini 5*'s motto, and a covered wagon was the motif of their crew patch. Joining Neil as the second member of the backup crew was Elliot See.

"I was really pleased to be assigned a mission and to fly with Elliot," Neil said. "Who wouldn't be pleased to back up Gordon Cooper and Pete Conrad?

"We became a very close team, and for months we spent almost all our time together getting ready. We spent a lot of time at the plant in Saint Louis working on and testing the spacecraft. So we all knew it very well by the time it was shipped to the Cape.

"We never lost sight of the fact everything was based on beating the Russians and getting there by the end of the decade," Neil emphasized. "That's why the schedule was overwhelmingly important.

"From the competence-building point of view, it was very good because the backup pilots essentially learned everything about that flight so they could not only take that information forward but when the flight was actually under way, they could be useful in Mission Control—not necessarily working as CapCom talking to the crew in flight, but being around and being available to talk to whomever wanted more information about how the crew flying did this thing or that, or whether it would be okay if Mission Control asked them to do this or that. So there was a very useful role for the backups."

As for his specific assignment backing up Gordon Cooper, Neil had nothing but praise for *Gemini 5*'s prime commander. "Gordon and I worked well together," Neil said flatly, pointing out that he had been assigned as a member of Cooper's support team for Mercury's final flight. He had become very impressed with the Oklahoma rancher's flying abilities—not because he was a small-town boy, too, but because Gordo Cooper had the flying thing nailed.

The problem was that some NASA bosses didn't care for Gordo's pranks, like putting live saltwater trout in Henri Landwirth's motel pool just so he could catch a fish, or driving race cars at the Daytona Speedway, or joining his friend 1960 Indianapolis 500 winner Jim Rathmann to outsmart and outrun the "Smokies" on Florida's highways.

Neil had heard, before the final Mercury flight had been set in concrete, that Operations Director Walt Williams had stopped by Deke's office and said, "Look, I know besides you Gordo Cooper is the only Mercury guy who hasn't flown. But maybe it would be a good idea to consider moving Al Shepard into this last Mercury flight." Then Williams saw Deke Slayton's face, and felt a definite chill. "Of course, it's your call, Deke."

Deke simmered as he nodded a weak good-bye to Williams.

Gordon Cooper was too much of a maverick for some bosses in the space agency. His hotshot jet flying and his tendency to bend the rules did not sit well with some. Deke judged Gordo as nothing less than a terrific pilot. He had come up through the ranks—paying his dues all along the way, flying everything from J-3 cubs to F-106s—and he belonged at the stick of the last Mercury. If anyone knew how it felt to have an earned mission yanked from under his feet, it sure as hell was Deke Slayton. He wasn't about to stand by and see Gordo get the shaft.

And there was something else that grated Neil Armstrong's sense of fair play.

There were these elitists who disapproved of Gordo's Oklahoma twang. "He's nothing but a redneck," laughed some members of the media and NASA's public affairs office. To them the fact that Leroy Gordon "Gordo" Cooper Jr. was one of the best pilots on Earth was irrelevant. They just didn't want "trailer park trash" representing America in space.

Well, as Neil had been told and witnessed for himself, Gordo Cooper met this problem as he did all of his problems: head-on. He grabbed the NASA public affairs officer leading the attack on his heritage and threw him into a hallway wall assuring him he would kick his condescending ass. The man's only defense was to "hide behind the rules and laws drafted by lesser men" and then to run. Scared out of his wits, the NASA mouthpiece rushed into Deke's office only to be told, "If Gordo needs any help kicking your ass I'll help."

That was the end of it. The flight was Gordo's and Neil was delighted. After 19 of 22 planned orbits on Cooper's Mercury flight most of his spacecraft's systems rolled over and died, and with only a single battery-supported radio and no power, ace pilot Gordo Cooper flew his ship to a landing within sight of the recovery ship. Neil Armstrong was first to say he knew of no other pilot who could have handled a dead spacecraft as Cooper had done.

Neil smiled. "If I ever fly combat again, I want Gordo Cooper on my wing."

The training for the *Gemini 5* crews was demanding. "The reality of the world in those days is that a lot of testing took place at two o'clock in the morning," Neil said. "And we were relieving each other when we could.

"The four of us spent enormous amounts of time together, working out the details.

"I would not say that we never cracked a joke or talked about something off the project. We were 98 percent focused on the job we had to do, but we did manage to get home, back to the neighborhood occasionally."

Neil would spend his rare downtime enjoying family, getting to know his toddler Mark and playing catch with Ricky, throwing a couple easy ones for a seven-year-old to hit, and when possible, he would try to steal a few hours for personal flying.

Neil not only flew machines with motors, he was also an elite glider pilot. He loved being alone high above the flat Texas landscape with only the sound of air rushing past his canopy—wafting on thermal currents, circling in wide arcs.

"You're just floating," Neil told me. "No engines. No artificial propellant. You're the hawk floating and twisting in the sky."

Neil was convinced that along with hang gliding sailplane gliding was about as close as a human could get to truly soaring like a bird, and he loved it. Even though he had flown the fastest jets ever, including the X-15 rocket plane, he still belonged to a local flying club. He still enjoyed slow, single-engine land aircraft as well as gliders.

One of Armstrong's flying club buddies was Bob Button, spokesman for the astronauts and a three-time-wounded combat veteran.

One early evening Button and Neil and Jack Riley, another NASA spokesperson, decided to take a Piper Tri-Pacer up for some night flying. "We wanted to bore some holes in the black sky, and I took the pilot's seat," Button said. Can you imagine that? Bob Button putting Neil Armstrong into the copilot's chair?

That's like him replacing Peyton Manning at quarterback. Riley went to sleep in the back.

The weather was CAVU (Ceilings and Visibility Unlimited) with none of those low cumulus clouds that loved to hover around Houston like fat moths, and Button rolled the Tri-Pacer onto the runway and made a smooth takeoff and climbed into blackness. As they passed through 5,000 feet the three were astonished to see the pure night absorb all light. They were suddenly over the Gulf of Mexico with only an occasional dim light from a shrimp boat below.

This was the peace the three sought, and they flew silently. No one said a word. They loved the remoteness from the hubbub of the city and welcomed the solitude that went on for about an hour until they could no longer see the lights of Houston.

Then Neil broke the silence. "Okay, let's head back."

By now, Button had the Tri-Pacer topping ten thousand feet and it was a long way down for the little single-engine land aircraft. No need to let carburetor ice build up and kill the engine Button reminded himself. Carburetor ice could be a problem for small aircraft coasting down from such altitudes, and he reached for the carburetor heat knob and pulled it out.

Silence!

Button had pulled the wrong knob! Apparently he pulled the mixture control, and starved the Tri-Pacer's engine of fuel.

In front of them the propeller slowed until the slipstream set it windmilling. The small plane responded, pitching its nose downward as it headed for the gulf, headed for the pit of blackness below; the sudden silence woke Jack Riley.

"Whaaat the hell," he yelled, springing up from the backseat so fast he almost put his head through the Tri-Pacer's fabric roof.

No one answered him. All Riley could see were Button's frantic hands moving controls, trying to get the fuel mixture back to full rich.

An unflappable Neil Armstrong sat quietly. He calmly watched Button deal with their impending problem of becoming a Gulf of Mexico submarine.

Braaaaahh!

There it was. They welcomed the engine suddenly coming to life again and powering the propeller. Button enriched his fuel mixture and shoved the throttle forward. The propeller bit hard into the gulf's moist air and as it spun faster and faster, the three souls on board were relieved to see the plane's nose come up and to feel the Tri-Pacer hauling them upstairs.

Within seconds the Tri-Pacer was leaving the dark water in its wake, gaining altitude, and they could again see the lights of Houston.

Neil Armstrong said not a word until Bob Button had all his flying duties under control.

Then not a scolding from one of NASA's most experienced test pilots. He only offered a constructive suggestion: "Bob you may wish to keep in mind what they teach in test-pilot school." Neil smiled. "When you change an airplane's control in flight, hang onto that control until the airplane does what you want."

Button didn't have to be run over by a truck. He immediately recognized what Neil was telling him: that keeping your hand on anything in life until you achieve the outcome you wish is the quickest way to complete a task. Had he held onto the wrong control he had moved, he could have immediately returned the control to its correct position without harm. Closing the barn door after the horse is out doesn't get the job done.

Cosmonaut Alexei Leonov takes humankind's first walk in space. (Russian Federal Space Agency)

EIGHT

THE GEMINI TWINS
ARE FLYING

March 18, 1965, America was caught with its pants down, again.

Alexei Leonov stepped into space linked to his Russian craft by a lifeline. He floated for twelve minutes to perform history's first Extra Vehicular Activity (EVA)—an ability necessary for building structures or reaching other places in space.

But despite being beaten again America wasn't standing still. The country was no longer stumbling. NASA's plans for reaching the moon and landing there were gathering steam. Project Mercury had been thoroughly successful. The first Gemini two-seater was on its pad, ready for launch in five days.

Neil Armstrong had been given a major role.

Deke Slayton had selected Gordon Cooper as the *Gemini 3* astronauts' primary capsule communicator—the CapCom. He was sending Armstrong to NASA's primary tracking station located in the country's southernmost state, Hawaii, on the island of Kauai. Neil would be sitting as CapCom in the middle of the Pacific, the perfect place to help *Gemini 3*'s crew Gus Grissom and John Young prepare for reentry across Earth's largest ocean.

It was a good assignment. Neil had been there before with a support group made up of the new astronauts for the final Mercury flight flown by Gordon Cooper. He had studied the Hawaiian tracking station's history. It was located on the most northern of the major Hawaiian Islands. The station was tucked away in the mountains of Hawaii's Garden Isle on the northwestern rim of Waimea Canyon, described by most as the "Grand Canyon of the Pacific."

The scenery was so breathtaking and relaxing Neil brought Janet along. It was a feel-good moment. Janet was a fort as a mother and wife—a daughter of a medical doctor educated strongly in her own right. Neil knew from the moment Janet agreed to marry him, he'd been blessed. Never once had Janet faltered. Never once did Neil hesitate to leave home and hearth in her hands. No matter where NASA sent him, having to worry about how Janet was handling the home fires was never a concern. A happy Janet telling Hawaii station director Virgil True, "This is my first vacation in seven years," brought a pleased smile to Neil's face.

God knew she deserved it.

Virgil True would serve four decades as director of the Hawaiian station. He and Neil became friends and Neil soon learned the station was only 5 degrees south of Cape Canaveral's orbital insertion of 28 degrees inclined with the equator. Its location permitted Hawaii to track more orbits than any other on NASA's worldwide tracking network.

For this reason Kauai was designated the "primary" station and transmitted verbal commands to orbiting spacecraft. "Secondary" stations handled radar and telemetry information, but like most in NASA Neil found it difficult to get his mind off of Alexei Leonov's first EVA. His thoughts kept going back to his friend and next-door neighbor Ed White.

White was assigned the right seat of *Gemini 4*. Jim McDivitt in command would fly the left. In two or three months they were to follow *Gemini 3* into orbit. Their flight plan called for them to pressurize their spacesuits and then depressurize *Gemini 4*. They would then open Ed White's hatch and he would stand up. This would expose most of his body and suit to space. That was to have been NASA's first step at mastering Extra Vehicular Activity.

Neil's best guess was that with the Russians beating them to the first EVA, NASA bosses would have Ed make a full spacewalk. Neil was aware of Chief Flight Director Chris Kraft's candor when he complimented the Russians, admitting to the public that the Extra Vehicular Activity "was a tremendous surprise." Neil knew the Russians clear warning to their cosmonauts that they were headed for the moon was well noted by his agency. He could visualize flight planners laying out a full EVA for Ed but he also knew he would not learn the real details until he was home and talking with Ed over their shared fence.

Within NASA there might have been even more concern about Russia reaching the moon had they known the details of their plans for cosmonauts Pavel Belyayev and Alexei Leonov—the crew that had just conducted the first EVA.

Soon after the cosmonauts return, mission commander Pavel Belyayev began training for a circumlunar flight, a loop around the moon, which would be made in about two-and-a-half years. Spacewalker Alexei Leonov was assigned to the group training to go to the moon and possibly land. In fact Leonov would become the commander of the Soviets' lunar-landing program and be promoted to the rank of general. He and cosmonaut Oleg Makarov were planning to follow Belyayev in Zond, a larger spacecraft currently being built. It would be the first two-man circumlunar flight.

The Russians were most aware of the propaganda value of flying around the lunar landscape first even if they couldn't land. They knew that if Belyayev could complete a flight sooner by doing it alone, then they could lay claim to and no doubt be given full credit for having reached the moon first.

The Soviets settled in with plans to one-up the Americans again as NASA moved through the countdown for its first Gemini. Neil waited in his Hawaiian CapCom chair. He wanted to be ready to help Gus and John when they came his way.

Gemini 3's countdown followed its script. The new spacecraft raced into orbit and began the first of three scheduled trips around Earth. The flight plan called for the crew to wring out the all new two-seater—shakedown all its systems. Gus and John did just that. They tested every one of Gemini's working parts including firing the first-ever group of onboard rocket thrusters. That changed their orbit. They flew maneuvers essential for going to the moon, for rendezvous, and for docking. They quickly came to the conclusion that flying the new two-seater was a delight.

Then, when Gus and John reached Hawaii, they heard a friendly voice. "Molly Brown, this is Hawaii CapCom."

Commander Gus Grissom had changed *Gemini 3*'s name to the Unsinkable Molly Brown after his first spacecraft *Liberty Bell 7* sank. He wasn't about to have a repeat on this mission. He cleared his throat and fired back, "Hello, Hawaii, this is Molly Brown, and Neil you're not going to believe how this baby handles. It flies just right."

"That's great, Gus. I can't wait until I get my turn," he said, before going into all the numbers and fuel levels they needed to keep their flight going. Then Neil told them, "Everything looks good on the ground. We'll see you next time around. Aloha."

"Aloha, Neil." Gus waved and *Gemini 3* swept across the Pacific, heading for its pass over the 48 contiguous states.

Gemini 3's maneuvers were bringing the moon just a bit closer, and when Gus and John reached Hawaii again, they would execute a final test-firing of their new spacecraft's rockets before setting up for reentry and splashdown.

But first *Gemini 3* had to complete its last trip around Earth. It raced across the other side of the world, down below sweeping over Australia before beginning its climb across the Pacific, taking dead aim for its final Hawaiian pass. Gus and John smiled when they heard the crackle of an incoming transmission. "Hello, Molly Brown. Hawaii CapCom."

"Hawaii, Molly Brown. All ready to burn," Gus told Neil, meaning the final firing of their maneuvering rockets was ready.

"Roger," Neil acknowledged. "We're right on. Hawaii has radar contact. Give us a start when you burn."

"Getting ready to fire," Gus came back. "Mark."

"We've got your start of the burn," Neil told him.

"That's good," Gus said, reporting, "There's 90, 70, 60, 50, 40, that's 20— 10 percent to go."

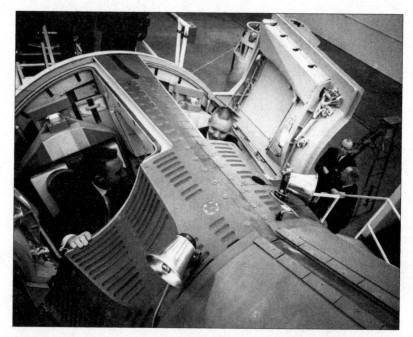

Gus Grissom and Neil Armstrong logging time in the Gemini simulator. (NASA)

"Hawaii command carrier off."

"Mark," Gus shouted. "End of burn."

"Good show," Neil shouted. "You looked good on the ground," he added quickly as Gus and John flew away and headed for a perfect reentry and landing.

With the success of Gemini's maiden launch, NASA tightened its schedule and kicked Project Gemini into high gear.

Deke Slayton told *Gemini 4* commander Jim McDivitt his crewmate Ed White would perform America's first full EVA. Ed and Jim went off with the experts and developed a plan.

Six weeks later, *Gemini 4*, the new Houston Mission Control Center's first flight, roared into orbit. On its fourth trip around Earth, Ed White opened his hatch and stepped out into his own orbit.

Neil, back in Houston, didn't want to miss a single second of Ed White and Jim McDivitt's four days in space. He didn't. He could see easily what was happening when Ed stepped outside over the blue Pacific gripping a handheld gun armed with pressurized oxygen. It fired in timed spurts; a steering jet that pushed him in the direction he sought and stopped his movement when needed.

Neil immediately knew the fun Ed was having and he couldn't get over the incredible view beneath his spacewalking friend.

Ed White appears to float on a mattress of puffy clouds. (NASA)

Ed White was beginning to understand his inability to move his own body when suddenly he was startled. The strangest satellite yet launched floated before him. It was a thermal glove he had left on his seat. It drifted up and away and began its own orbit, and Ed scolded himself with one word: "careless."

Then, not at all aware the twelve minutes planned for his outside adventure had quickly passed, it was time to reenter *Gemini 4*. He was running out of daylight, but he was having too much fun.

Gus Grissom was CapCom in the new Mission Control Center 27 miles southeast of Houston, and he knew the euphoria White was showing was akin to the dangerous "raptures of the deep."

Grissom, in the veteran Mercury astronaut's deep command voice, told the frolicking spacewalker, "*Gemini 4*, get back in." Neither Grissom nor White was aware of how their fates were so intertwined.

Ed White's crewmate and commander Jim McDivitt repeated the order: "They want you to get back in now."

The fun-loving spacewalker still hesitated. "What does the flight director say?" he asked.

"The flight director says get back in," barked Chris Kraft.

"This is fun! I don't want to come back in," White laughed, performing one more somersault before telling Mission Control, "Okay, I'm coming in."

Ed reluctantly began his return to *Gemini 4* but quickly discovered just to move his body even one inch from its independent orbit without help wasn't at all easy. He needed an assist. He needed a jet backpack or a line to pull. Otherwise he'd simply drain his energy and his body would begin to overheat.

Ed White made it back in but he'd just discovered a major difficulty that would confound future spacewalkers.

Gemini 4 returned after four days of orbiting Earth, and that August, *Gemini 5*, with Gordon Cooper and Pete Conrad, doubled *4*'s record by spending eight days in space. Their backups Neil Armstrong and Elliot See were at the Cape for *5*'s launch, and immediately left for a quick flight in their T-38 to Houston's Mission Control.

Armstrong and See spent *Gemini 5*'s eight days in Mission Control in the comfort of air-conditioning, hot food, and ready refreshments while Cooper and Conrad spent those same eight days inside their slightly-larger-than-a-phone-booth spacecraft. They proved humans could function long enough in space to fly to the moon and back, and when the *Gemini 5* astronauts returned, newspapers ran a cartoon of them holding hands on the recovery ship's deck with the caption: "We're engaged."

Gemini 5 was one of those fun flights and three weeks after its splashdown Deke Slayton made it official: He cornered Neil Armstrong and told him he would command *Gemini 8*.

Neil nodded, smiling.

Deke then told him his crew member would be Dave Scott, a West Point graduate and the first member of the third group of astronauts to fly.

Deke got no argument from Neil as the next group of astronauts, a small assemblage of six scientists, were named; Owen Garriott, Edward Gibson, Duane

Graveline, Joseph Kerwin, Curt Michel, and Harrison Schmitt. Schmitt would become the only scientist to reach the moon.

Astronauts Frank Borman and James Lovell bettered *Gemini 5*'s record by spending two weeks inside *Gemini 7*'s cramped quarters, and Lovell would later say, "It was like spending two weeks in a men's room."

The monotony of Borman and Lovell's marathon mission was broken on day eleven by visitors from Earth.

Wally Schirra and Tom Stafford were flying *Gemini 6*. They rode their two-seater right up to *Gemini 7* and put on the brakes. They thought they would

Gemini 6 and *Gemini 7* meet in space. (NASA)

stay for a while and Schirra and Stafford began "station keeping" with Borman and Lovell. It was the first rendezvous in space.

"We've got company," Lovell reported.

"There's a lot of traffic up here," Schirra quipped.

"Call a traffic cop," Borman laughed.

American astronauts had used the Gemini spacecraft's rockets to perform a meaningful first and the two Gemini ships together orbited Earth in formation, doing fly-arounds and circling each other in a series of figure eights. Wally Schirra told Mission Control he closed to a distance of six to eight inches, backed off, and flew in again.

Schirra was the only member of the Mercury Seven "pranksters" in the

Neil tries on his *Gemini 8* commander's spacesuit. It fits. (NASA)

group, and when he flew his *Gemini 6* away from *Gemini 7*, he and Tom Stafford shook up Mission Control when Stafford reported he had seen a UFO.

It was only days before Christmas and when startled flight controllers questioned Stafford about the Unidentified Flying Object he had reported, his answer was calm in his casual Oklahoma monotone: "Well, it appears to me, Houston," he began explaining as Schirra was heard in the background playing "Jingle Bells" on his harmonica, "it's some sort of a sleigh. It's coming down from the North Pole with a jolly old man dressed in a red suit driving a bunch of reindeer. He seems to be headed your way, Houston. Better tell all the children old Saint Nick is on his way."

This broadcast reporter took the *Gemini 6* "Jingle Bell's" UFO tape and did a full Christmas Eve report on all NBC networks as America enjoyed the holidays. With the rendezvous of the two spaceships, another milestone had been reached in the country's road to the moon. But no one had docked ships in space. That little job was still out there. It now fell to Neil Armstrong and Dave Scott.

The *Gemini 8* crew took a couple of days off, and ate a little Christmas dinner before getting back to their 24-7 training.

The rendezvous between *Gemini 6* and 7 had not been in NASA's original plans. Schirra and Stafford were originally to have chased down and docked with an unmanned Agena rocket. But the Agena blew up on its way to orbit, and agency officials came up with the ingenious plan to launch *Gemini 7*'s two-week mission first, and then send *Gemini 6* in pursuit.

Now the next step was up to Neil Armstrong and Dave Scott. They loved the challenge. They spent a great portion of their training hours in the docking simulator and Neil felt the simulator was a good representation of what they could expect.

"Dave and I thought of our mission as being an absolutely super flight with great objectives," he said. "And with Dave Scott in the right seat," he added, "I believed we wouldn't have any trouble."

Neil could not have been more pleased with his crewmate. He'd known Dave at Edwards, but not well. He had come to know him better in Houston and found him very likeable and good at what he did.

"Dave was diligent and he was hard working," Neil said. "I felt confident in his ability to handle his part of the responsibilities, and another plus, I had spent much time learning about the Gemini spacecraft as backup commander for *Gemini 5.*

"The differences were those things that were simply different," Armstrong added with a smile. They were going to rendezvous and dock with a "live" Agena rocket and Dave Scott had an extravehicular backpack in the back of the spacecraft for a longer spacewalk than Ed White made, and they had more experiments.

The *Gemini 8* crew had their training in the bank by tax day. They were confident they were ready.

Dave Scott and Neil Armstrong stand by to board *Gemini 8* as they laugh at pad leader Guenter Wendt's latest joke. (NASA)

NINE

GEMINI EIGHT

Neil Armstrong and Dave Scott went over the seals of the *Gemini 8*'s hatches at 8:38 A.M. eastern time March 16, 1966.

Their cramped two-seater spacecraft would be their home for the next three days so they got busy hooking themselves up to health monitors and opening communications with Mission Control while making sure everything they needed was on board. They would be sleeping in their suits in their seats. No shower; only wet towels. They double-checked to make sure they at least had their toothbrushes.

At the same moment on the nearby Atlas/Agena pad, the countdown for launching their target rocket was going well. They were told they would be able to see the Atlas/Agena lift off while back in Houston; outside Neil's home Janet was busy circling the wagons.

An assemblage of photographers, television crews, broadcasters, newspaper writers—all sorts of media members—had gathered and were playing the waiting game. They were hopeful Janet would emerge to talk with them, tell them how she felt, what her emotions were. Were they everything from pride to fear . . . fear that Neil might be killed?

This wasn't Janet Armstrong's first rodeo. She'd waited before when Neil had flown the X-15, when he had flown crucial tests in high-speed jets. And now she was waiting again, telling herself this is what she'd signed up for. This was the role of a test pilot's wife and she had freely chosen this role, and she would do what was expected of her.

She would stay inside their home and wait for the launch, dealing privately with her own nervousness. She and their boys Rick and Mark would watch television and listen to the NASA squawk box. The agency had installed it for her so they could hear and cheer every word between Mission Control and their husband and father.

Knowing his family was secure for his and Dave's three-day flight without having to fight the crowds and the media at the launch site put Neil at ease. But his mom, dad, sister, brother, and other family members were another thing. They were at the moment riding a family bus along with some of Dave's family and relatives and they were singing and very excited. They were headed to family bleachers to view the launch when suddenly they were startled.

It was 10:00 A.M. eastern time and the Atlas/Agena rocket rose from its launchpad on what seemed to be a never-ending column of fire, and the driver quickly stopped the bus.

Cheering broke out as some leapt through the bus door. They shouted loudly, willing the *Gemini 8*'s target rocket on to success.

They watched with fingers crossed until the Atlas/Agena's fiery contrail became only a pinpoint of light in the Florida sky.

NASA commentator Paul Haney in Mission Control reported, "When Neil Armstrong heard that the Agena had ignited and was performing well, he was told, 'It looks like we have a live one up there for you.'

"Neil came back with, 'Good show.'"

In their seats, Neil and Dave were ready. Their target for rendezvous and docking was now in an orbit 161 by 156 nautical miles, as close to a circular orbit as they could hope for with an unmanned vehicle.

Gemini 8's Atlas/Agena target rocket launches. (NASA)

"Beautiful, we'll take that one," Neil told Dave.

"You betcha," his partner agreed.

With the Atlas/Agena launching on time their mission had a great start. Now if we can only be as precise with our launch, Neil reasoned, it would mean we could fly our mission the way we had practiced it in the simulator.

So far the wee ones were smiling and the countdown was on a track not to disappoint.

There had been only one minor problem. Some epoxy had glued shut one of Dave's harness mechanisms and Neil had called in John Wayne and the cavalry. Astronaut Pete Conrad, his backup, and pad leader Guenter Wendt came to the white room and rubbed and cleaned the mechanism until they got the catch unglued. From then on, the countdown had been smooth sailing

In their seats and ready. Dave and Neil are strapped in to fly. (NASA)

with the crew checking the positions of all the switches on their consoles, constantly reporting their spacecraft's condition to Mission Control, static testing their Gemini's thruster rockets, and assuring the flight surgeons they, too, were ready.

Suddenly they found themselves in a 5-minute, 54-second hold at T-minus three minutes. This delay in their countdown was to bring their orbital insertion in line with their Agena target.

They were almost there and Neil instinctively checked his harness once more as he heard the launch director say, "We are resuming the count; T-minus three minutes and counting."

Neil lay back and locked his spurs into the bottom of his seat. He and

Dave listened to launch controllers from one console to the other declaring they were "*Go!*"

"Ignition is at 40 minutes and 59 seconds past the hour," CapCom Jim Lovell told them, and Neil felt the Titan II coming alive through subtle groans and creaks. Fluids were shifting. Pressures were increasing. Pumps were being tested. The ten-story-tall rocket beneath was making itself ready to push *Gemini 8* into orbit at speeds greater than 17,000 miles per hour.

"This is the launch director. We have a clearance for launch at T-minus one minute and forty-seven seconds and counting."

Neil shot a look at Dave. He was grinning.

Inside the Armstrong household, family and friends stared at a television image of the rocket with *Gemini 8*, and hugged their NASA squawk box listening intently to NASA commentator Jack King in launch control:

> Now coming up on T-minus 90 seconds, mark T-minus 90 seconds and counting. All systems looking good during this final phase of the countdown, and at zero, the two engines will ignite and build up some 430,000 pounds of thrust just prior to liftoff. Once the vehicle builds up 77 percent of this thrust we'll get a go for liftoff. This will occur some four seconds after ignition. . . .
>
> Now T-minus one minute and counting as we go through our final checks; T-minus 50 seconds and counting, if all goes well *Gemini 8* will be inserted into orbit some 1,015 nautical miles behind the Agena; T-minus 40 seconds and counting . . .
>
> In the blockhouse the crew is reporting as they monitor the various activities over the final phase. Now T-minus 30 seconds and counting; T-minus 20 seconds and counting . . .
>
> 15 . . .
>
> 5, 4, 3, 2, 1, we have ignition.

No one needed to tell Neil Armstrong his rocket was alive. Flame and thunder rolled over the pad and he and Dave felt its explosive fury. Four hundred and thirty thousand pounds of thrust from its first stage was straining against its huge hold-down bolts as its two engines built to full power. Three seconds later, Neil heard rifle shots crack through the thunder and he knew the hold-down bolts had been sheared. Immediately he felt movement.

They were on their way. The only surprise? Neil couldn't believe how smooth their liftoff was and he heard CapCom Jim Lovell tell him, "Good liftoff, *8*."

"Roger," Neil responded.

They were on the flight director's audio loop and Neil immediately felt the Titan roll. He and Dave were now headed into orbit on their side and he told Houston Mission Control, "Roll program is in and pitch program is in at 30 seconds."

Neil feels the rocket beneath him come to life. (NASA)

"Roger, pitch program," Lovell answered, adding, "Mark. Fifty seconds. You are looking good, *8*."

"Roger," said Neil as they continued gaining speed, and he stared straight up through his cat-eye window at the bluest of Florida skies. Already they were

Gemini 8 heads into space. (NASA)

pulling a G load and he was being pushed deeper and deeper into his seat as he watched feathery white clouds scatter as if they were being chased away by the thundering Titan II pushing him and Dave upward and upward. All at once he knew they were punching through Max-Q, a sudden intrusion of the atmosphere's strongest pressures into their smooth flight, which reminded him of when he was 12 or 13 riding his old bicycle without fenders down that rutted road, bouncing along to his job at the local airport.

His rocket was now shaking and twisting his limbs as did that old bicycle, and he managed a quick laugh remembering how his dog Skip would run along barking at everything and getting just enough in his way to make him jerk the handlebars even more.

He loved that dog. He loved him as long as he stayed out of the briars and he didn't have to dig through the thorns that would rip into his arms and hands to untangle Skip's paws and legs.

That was pure pleasure then and this is pure pleasure now being shaken by the Titan II. It was fighting its way through squeezing forces and Neil listened to the engine roar and high-pitched howls of air ripping past vanish with the supersonic speeds as the sky turned from rich blue to black. Neil was suddenly aware this was where he had been in his X-15 in April 1962.

He had topped nearly forty miles that day and he recognized the place and—"*Gemini 8*, this is Houston." Jim Lovell interrupted Neil's thoughts. "You are *Go* from the ground for staging."

"Roger," Neil acknowledged.

The two astronauts were ready. Suddenly a sheet of fire washed over their spacecraft with a heavy jolt. For just an instant it could have been bad news, but both pilots recognized their second stage had ignited. They knew the second stage fired while still attached to the first, and the sudden blow of 100,000 pounds of flaming thrust was what ripped the two stages apart.

It was not at all pleasant and they caught their collective breaths and waited for full pitch over, waited for *Gemini 8*'s nose to come down, and suddenly Neil could only see black sky.

"I understand you have guidance?"

"Rog, we have guidance," Neil told Mission Control. "Zero pitch and one degree yaw right coming in."

"Roger, your guidance looks good on the ground."

"Pitch . . . yaw about a quarter of a degree," Neil told Lovell.

"Roger, your plots are looking very nominal here on the ground, *8*."

"The second stage was real good, Mission Control."

"*Gemini 8*," Jim Lovell told them, "you are *Go* from the ground."

Neil Armstrong and Dave Scott were now relaxed and enjoying their ride. They were ahead of the curve on all their flight duties and their spacecraft nose had pitched over giving them an unbelievable view of the horizon, and they drank it all in as they looked down on the world from their seats with the gods, the sky before them the blackest of black with a thin band of blue and whites along the horizon.

They watched as the waters of the Atlantic covered by layers of clouds slid beneath them and they heard Jim Lovell tell them, "*8*, you have reached 80 percent of the velocity desired. You are now 85 nautical miles high and looking good."

Neil and Dave looked at one another with broad smiles and nods and began their procedures for second-stage cutoff and separation followed by the first firing of *Gemini 8*'s own rocket thrusters. Neil would fire the thrusters. Dave would feed him the numbers. They needed a little more push from the spacecraft itself to reach the orbit they needed to chase down and dock with the Agena target.

The Armstrong household, as did millions viewing the flight of *Gemini 8* on television, heard NASA commentator Paul Haney report:

> We have second-stage cutoff, approximately 5 minutes, 40 seconds (*into the flight*). Five minutes 50 seconds, in about 10 seconds the crew should initiate their thrusters. Flight dynamics confirms again he is go and Lovell is passing this up to the crew. Six minutes, 5 seconds into the flight, and Armstrong advises they have completed their burn (*rocket thrust firing*), and they are free of the second stage. Six minutes, 40 seconds into the flight and as yet we've heard no numbers on the orbit, but we believe it will be very close to the planned value.

The miracle was real. Two flawless countdowns followed by two perfect launchings—all in one day from the sands of Cape Canaveral, and Neil Armstrong and Dave Scott were orbiting Earth at 17,500 miles per hour—entering their orbit at 87 by 147 nautical miles, its lowest and highest altitudes. This was comforting for flight controllers who had sought an orbit of 87 by 146, and *Gemini 8*'s chase of its docking target had begun.

The astronauts shot each other a thumbs-up, and Neil felt as if he was a stranger in this new world. He wasn't sitting up or lying down in this unbelievable place where up and down no longer existed. He was floating in physical limbo with only his harness holding him in his seat as loose objects drifted in *Gemini 8*'s cabin while before him, looking out his cat-eye window, he could tell the sun behind him was low above its horizon.

Neil could now see the western coastline of North Africa where clouds obscured most of it to Casablanca. From there, he could see landmarks all the way to the rock of Gibraltar and out farther before them the sculpted sands of the great Sahara. The sun reflecting low across the desert told him he was about to see his first night in space.

Neil and Dave enjoy their first view of orbit. (NASA)

The twilight was simply beautiful, and for the next five or six minutes Neil watched the slow but continuous reduction in light intensity in the brilliant orange-and-blue layers that appeared to be blankets protecting Earth beneath the totality of the universe's black sky. And then, 34 minutes into their flight, all light dimmed into the completeness of dark, and night was upon them. They could suddenly see brilliant fire streaming from *Gemini 8*'s thruster rockets as they fired; but it was not the only fire. Far below, thunderstorms were spitting snarling fiery bolts that created their own jagged, dancing patterns, and Neil turned down the lights in their spacecraft's cabin to further acclimate themselves to the darkness.

The small towner was simply amazed, as he had been in childhood, seeing so many stars. They were more rich and defined and the Milky Way was—hell it was the Milky Way—unbelievable. It was so defined and definite wrapping itself around them as Earthglow, even from the dark side, gleamed softly along their spacecraft with light from forest fires and burning plumes of oil and natural gas wells from Lagos, Nigeria, below. In the atmosphere itself, continuous lightning danced along its clouds, staying with them as streaks from meteors burned fierce paths through Earth's protective layers of dense air.

Neil stares in wonder at the Milky Way—so distinct and breathtaking. (NASA)

Above Neil and Dave was their solar system, their masterly detailed Milky Way, and beyond were great galaxies, dense clusters, and black holes. Along their edges were exploding stars being crushed by the universe's most powerful gravity, and beyond—yes beyond—was the infant universe ten, eleven, twelve billion light years away where only the clearest and blackest skies could be found. When they returned their eyes to their own solar system—most notable to Neil from his astronomy days—were the planets, just hanging there like distant lighthouses in the constant dark.

Neil Armstrong would never admit it to a living soul, but the gods were truly welcoming him to their beautiful home.

A view of the Hawaiian Islands from space. (NASA)

TEN

GEMINI 8: THE DOCKING

Once in orbit, the *Gemini 8* astronauts were following a demanding set of flight maneuvers to rendezvous with their Agena target. There was little time for sightseeing, but Neil Armstrong didn't want to miss the Hawaiian Islands as well as his old stomping grounds at Edwards. He and Dave Scott both wanted to steal enough time to see their homes near Houston, and at 1 hour and 12 minutes following their launch, they were making their first pass over the fiftieth state with Neil looking down to see Molokai, Maui, and the Big Island. But when Oahu came into view it was covered with rain clouds all the way to the tracking station on Kauai.

He smiled. "Hawaii this is *Gemini 8*," he called. "We have you in sight. It looks like a nice day."

"It's beautiful weather here, Neil, except for our daily scheduled shower."

"Is that you, Virgil?"

"Roger."

"As they say, Virgil, it's always raining somewhere in Hawaii."

"Roger that."

"Can't stop now, but we'll see you soon."

"We'll be here, Neil."

"Let's go with the update," Neil requested, and he and Dave grabbed their numbers and sped away. A five-second critical firing with the spacecraft's forward thrusters was first up.

The burning of their thrusters would keep them closing in on their Agena target and no sooner had they locked the numbers in the computer Dave Scott looked down and shouted, "We're going over the Los Angeles area now. Can you believe it?"

God, they both were green at this spaceflight stuff, Neil thought laughing to himself. They were a couple of giggling tourists in Hawaiian shirts and he simply could not contain his own excitement.

"Hey look at all those ships!" he shouted before moving his eyes inland. He was reasonably sure he could see Rogers Dry Lake bed; but could he see the cabin? There . . . that's where it should be . . . seven years he had spent in the high desert . . . memories . . . so many, but now wasn't the time for reminiscing.

He turned back to his duties and in less than seven minutes they were over the Houston area. He and Dave struggled to see their homes—too many clouds; not enough time to look.

It was 1 hour and 34 minutes into their flight. It was time for their next burn and Neil fired *Gemini 8*'s thrusters. They burned for five seconds, and the spacecraft's orbital speed decreased slightly. But there was suddenly another problem! When the firing was completed, it took a few seconds for the residual thrust to clear, which varied the computer readings making it difficult to tell the exact change in orbital speed.

The computer people on the ground went off to work on the problem while the crew, for the next 47 minutes, operated *Gemini 8*'s computer in "catch-up mode." Their duty was to continue the chase. *Gemini 8* was now below their Agena. That meant their spacecraft was traveling a shorter distance than their target, moving faster around Earth.

Neil and Dave were trying to lock-on its radar with the Agena's transponder. The transponder had one job—answer the inquiring signal from *Gemini*.

The terminal phase could not begin until they had a solid radar lock, and at 3 hours, 48 minutes, and 51 seconds into the flight, Mission Control was standing by to talk to the crew over the Tananarive tracking station.

Tananarive was located in the Malagasy Republic on the island of Madagascar off the east coast of Africa, and astronaut Jim Lovell raised the astronauts asking, "Do you have solid radar lock on the Agena?"

"That is affirmative." Neil delivered the good news. "We have solid radar lock. Just a second and I will give you our current range."

"Roger." Lovell smiled.

"We are indicating a 158-mile range and an elevation of about 4 degrees."

"Roger, sounds good and I'm seeing about three minutes and nine seconds until your burn."

Over Tananarive Neil Armstrong was ready to fire his thrusters again. The orbit they had been moving through to catch up with its target was elliptical, and Armstrong nosed down his spacecraft and triggered his rockets. The firing circularized their orbit in a more precise plane with the Agena, and *Gemini 8* continued to close on its target. Neil and Dave had radar giving them range and position and they knew they would see the Agena at some point, but just when and where?

Then at 4 hours and 48 minutes into the mission Neil told them to stand by: "Okay, we've got a visual on the Agena at 76 miles," he reported. "At least we have some object in sight. Or something that looks like it could be the Agena."

"Understand a visual Agena, or Sirius, 76 miles," acknowledged Houston.

"Yeah," Neil replied. "It could be a planet."

They moved ahead. Not sure they were seeing the Agena, but sure it was out there. They wanted to be in the dark for most of their approach until roughly ten miles out. That's when Agena would reenter daylight and they could see it like a huge beacon against the black of the universe and Neil called: "Okay, we could have a solid visual on the Agena," he told them. "Range is 56 miles. I'm taking a second look at it with the sextant."

Then, silence. Fifteen minutes passed with no word from the crew. Some flight controllers were getting edgy but spacecraft communicator Jim Lovell was hesitant to contact the astronauts because this was one of the most difficult parts of the flight.

"Both crewmen are quite busy," Lovell explained. "Dave is doing the math for the final approach and Neil is controlling the thrusters."

No sooner than Lovell had concluded his report, Neil was back. "Okay, going into darkness at—let's see, 05:02. We lost a visual on the Agena, and I had the ACQ [acquisition] light up right away. Range was forty-five miles." He paused. "It's very hard to see, but it looks like a sixth-magnitude star, I'd say."

Flight controllers looked at one another as *Gemini 8* moved deep into the night side of Earth. The astronauts' rendezvous and docking mission so far was pretty much peaches and cream.

Neil and Dave's hunt for their Agena quarry had gone better than most would have predicted and, with the distance between the two craft closing fast, they set up for their final push.

"Okay, I'm going to start braking down a little bit very shortly," Neil told Dave. "Are we inside 15K?" he asked.

"Inside 15K," Dave assured him (15K is 15,000 feet or 2.84 nautical miles).

The *Coastal Sentry Quebec* tracking ship below told Mission Control *Gemini 8*'s range from its Agena target was now about 14,000 feet and several flight controllers raised their thumbs indicating that things were going along very well.

Agena's light! It was straight ahead. Neil and Dave were fixed on its glare and Neil Armstrong kept his spacecraft closing the distance between the two at the glacial pace of about five feet per second, a speed any runner could best.

It was *the* critical point of the mission. Agena was so close. No time to lose concentration. Move in slowly and surely; put on the brakes at the precise time for station-keeping.

The maneuvers were at their most critical and demanding, but that didn't stop Houston from asking for more talk from the crew. "The pilots are acting extremely 'ho-hum,'" Mission Control told the *Coastal Sentry*. "Could you urge them to say a bit more about their situation?"

They were back on the day side, but Neil and Dave weren't talking, they were concentrating. They were making ever so sure they were crossing all their Ts and dotting all their Is, since in orbit they had been very much alone in their work. They weren't about to screw things up now by losing focus in banter with the ground. Not when . . . Suddenly, they were no longer alone!

It was right there!

Sitting right in front of *Gemini 8* was Agena. It was big. It was bold. It was beautiful. It was a product of Earth. Neil smiled at Dave. "That's just unbelievable," he said. "Unbelievable!"

"I can't believe it, either," Scott agreed, hitting Armstrong on the shoulder and saying, "Outstanding job, Coach!"

"It takes two to tango," Neil assured Dave.

The Agena target a football field away. (NASA)

They had completed their fourth run through the night side and they could now see a crack in the darkness followed by a breath of light that grew rapidly into a riotous shout of color, a vivid, glowing day.

The sun stabbed across half of the sleek and long Agena rocket stage. It was suspended in the universe's total blackness. Earthglow reached up with its warm blues and whites to welcome Neil and Dave. The crew of *Gemini 8* felt tremendous pride in the thousands who had built and launched the Gemini spacecraft and the Agena—who had sent them on this hunt and chase. They

Neil and Dave's Agena target 150 feet away. (NASA)

were grateful to, and honored by all those hands at the Cape, in Mission Control, in the Agena and Gemini plants that had made it possible.

But the biggest prize was still out there—the docking—and suddenly they heard Hawaii calling: "*Gemini 8*, this is Hawaii CapCom."

Dave Scott, knowing the hard work Neil had put into pulling off the rendezvous, smiled and said, "You tell them."

Armstrong nodded. "Aloha, Hawaii, we're station-keeping."

A cheer came up from the station and Neil continued, telling Hawaii and Jim Lovell in Mission Control that the Agena looked fine, the antennas were all in the proper position, and TDA (The Docking Adapter) hadn't a scratch, with no apparent worse for wear. They were in good shape for docking.

The two astronauts were pleased to have Jim Lovell as CapCom. He, along with Frank Borman, had flown the last Gemini three months ago—logged

Station-keeping with a perfect Agena shining in the glow of the Pacific. (NASA)

two weeks aboard *Gemini 7*, and Lovell certainly had solid knowledge of the ship they were flying.

They flew away from Hawaii again, but this time they had a wingman. They would fly in formation with the Agena across the Pacific and South America and on the other side of the continent they would find their next contact with Earth, the tracking ship *Rose Knot Victor.*

NASA commentator Paul Haney told the public and the media, "The flight plan from here calls for the actual docking to take place over the *Rose Knot Victor* parked down off the coast of South America. The *Rose Knot Victor* is to acquire the spacecraft at 6 hours 32 minutes into the flight. We presently show 6 hours 8 minutes into the flight."

Twenty minutes. That's what he had and Neil told himself to review learned procedures, to review all he had been told by Wally Schirra and Tom Stafford who, only three months earlier, had flown the first rendezvous, making up much of it as they flew *Gemini 6* right up to *Gemini 7*—putting on the brakes, station-keeping, doing fly-arounds, circling each other in a series of figure eights, and closing to within six to eight inches before backing off and flying in again. Now it was Dave Scott and Neil Armstrong's assignment to finish the job.

Were they ready? You betcha!

Neil and Dave began inching toward their target. *Gemini 8* had already moved to within 80 feet of the Agena and as they sped across the Pacific, Neil Armstrong fired his rocket thrusters as Dave Scott chanted distances and elevations. Neil's firing of the thrusters closed the distance between his spacecraft and the passive Agena an inch at a time. Closer and closer they came to the huge docking adapter on the back end of their target rocket, lining up perfectly with the funnel-shaped adapter. When they were within 2 feet, Neil put on the brakes. He matched *Gemini 8*'s speed and position precisely with the Agena's. They were in every sense one. They just hadn't docked.

Gemini 8's astronauts were peering down their spacecraft's nose into the docking ring. The sensors and latches that would secure their docking waited. Then they heard, *"Gemini 8* this is *Rose Knot Victor."*

". . . about two feet out," Neil told the tracking ship quietly.

"Roger, stand by for a couple of minutes here."

"Is he docked?" Mission Control asked.

"Negative, he's not docked yet," *Rose Knot Victor* told Houston, and then the tracking ship monitored and studied all the Gemini and Agena systems.

"Okay *Gemini 8*, it looks good here from the ground. We're showing column rigid, everything looks good for docking."

"Roger."

Neil Armstrong slowly nudged *Gemini 8*'s nose into Agena's adapter. Sensors recognized the spacecraft. The crew heard and felt electric motors grinding. The latches clicked home. *Gemini 8* and the Agena were one.

"Flight," Neil spoke directly to Flight Director John Hodge. "We are docked! It was really a smoothie."

In Houston Mission Control pandemonium broke loose: backslaps, handshakes, cheering, whistling, shouting, applause, and tremendous grins as cigars were given out by Deke Slayton. The control center flashed the news across

Gemini 8 aligned and ready for docking. (NASA)

NASA's tracking network, and throughout most of the world. American space flyers had achieved the first docking between two spacecraft.

"You couldn't have the thrill down there we have up here," Dave Scott told them, adding, "Just for your information, the Agena was very stable, and at the present we are having no noticeable oscillations at all."

NASA commentator Paul Haney was back:

This is Gemini Control Houston here at 6 hours, 44 minutes into the flight. According to the best estimates that we have the actual docking took place at 6 hours, 34 minutes into the flight. We can verify this later through telemetry but that is the estimate of the flight director. The pilots still have about two hours work ahead of them before they will

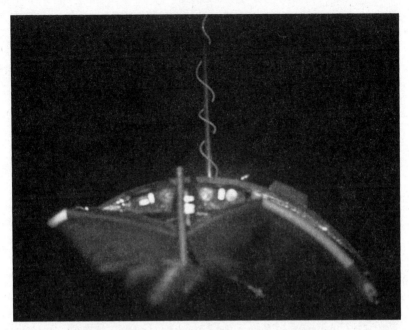

Neil Armstrong performs the first docking in space. (NASA)

power down for the night and suspend their activities after this most busy and successful day. This is Gemini Control, Houston.

Back on *Gemini 8* the crew and the docked ships moved out of radio contact with Mission Control. Ahead was one of the worldwide tracking network's dead zones. No transmissions in. No transmissions out.

For the next twenty-one minutes *Gemini 8* and its Agena partner, forming a single, lengthy spacecraft of 44 feet, moved across the waters of the Indian Ocean and over the Bay of Bengal where something went wrong.

There was no way of telling anyone. It was Neil Armstrong and Dave Scott's problem.

Suddenly, survival was job one.

Mission Control Houston ready for the quiet, graveyard shift. (NASA)

ELEVEN

GEMINI 8: THE EMERGENCY

I t was quiet on NASA's worldwide space-tracking network. Neil Armstrong and Dave Scott, with the successful first-ever docking of two spacecraft in their pockets, were winding down their day when Bill Anders walked into Mission Control. It was Anders's first real assignment as an astronaut—relieving his veteran colleague Jim Lovell.

Anders had come aboard in NASA's third group of astronauts. He had strategized by lobbying for CapCom; the experience would put him a step closer to a crew assignment. He hadn't the slightest clue he would, in less than three years, join Jim Lovell and Frank Borman as one of three humans to first orbit the moon.

At the present, *Gemini 8* was out of radio contact somewhere over the other side of the world, and Lovell, the flight's first-shift CapCom, told the rookie, "You should have an easy shift. The crew is getting ready to call it a night."

Anders smiled as the tired veteran astronaut turned and walked away, leaving him to his debut as a space-crew babysitter.

He now stood alone before the CapCom console surveying the rows of flight controllers, each with a monitor, rising behind and above the others, each tier keeping a continuous watch on the crew and spacecraft. This efficient Mission Control operation was orchestrated by the flight director, the boss, on the CapCom's right.

The freshman astronaut acknowledged Flight Director John Hodge with a glance and a smile. Hodge was also new. He was completing his first shift and his return smile seemed to say to Anders that he, too, would make it.

Hodge was about to be relieved by the more-experienced Gene Kranz, who in little more than three years would be the flight director in charge of Neil and Buzz Aldrin's first landing on the moon. Kranz had dropped by to listen to the docking, and since Hodge had been at the flight director's console for eleven hours, the two decided between themselves Kranz's second shift of flight controllers should report for duty immediately.

Rookie Bill Anders and the second shift settled into their Mission Control seats. Many from the first team hung around to make sure no one had a problem moving into the flow.

Over China, the docked *Gemini 8*/Agena combination flew alone deep into the night. They had no ground stations to talk to and Neil readied the spacecraft for their first sleep period. Dave continued maneuvering the Agena with its attitude-control rockets. He turned the docked ships 90 degrees to the right. This yaw maneuver was one of several that had been scheduled to determine if Agena's control system could relieve *Gemini 8* of some of its control duties. If it could it would save fuel. But this 90-degree turn took 5 seconds less than the full minute expected. This shorter maneuver prompted Dave to check his cockpit's instruments. Something was wrong. Something was out of place. The Gemini/Agena's stability was fraying noticeably. "Neil, we're in a bank," he said. "I have a 30-degree roll on my 8-ball."

The 8-ball instrument on a pilot's cockpit panel is an instrument of crosshairs showing a flyer an artificial horizon and the spacecraft's degree of tilt to the left or right "bank." It gets its name from the eight ball in the game of pool. It is held in place by a fast-spinning gyro, a device that holds an object fixed in place, and if the gyro fails, the 8-ball will tumble—no longer giving a pilot an artificial horizon and an indication of the spacecraft's bank. Armstrong quickly thought Scott's 8-ball had tumbled, but when he checked his own they were identical. His 8-ball showed them rolling and he made some efforts to reduce the bank angle by triggering short bursts on the Gemini's maneuvering thrusters.

At first it seemed Neil was successful. Both pilots were convinced their target rocket was the culprit because they could easily hear the Gemini thrusters when they fired. They were on the Gemini's nose and behind them and every time they would fire Neil would later say, "It was like a popgun—*crack, crack, crack*—and we weren't hearing anything, not realizing until later that the thrusters only popped when they ignited. As long as they were burning they were silent."

What Neil and Dave hadn't known was that if a thruster was burning continuously, they wouldn't be able to hear it. So they waited for four minutes. There was "no joy."

The roll began again, only faster, and Neil asked Dave to shut off all of Agena's controls.

Dave had the control panel for Agena on his side but he had little success. He sent command after command, jiggled switches, recycled them and the roll kept increasing and Neil became concerned that the stress and strain of the violent rolling might break apart the linked spacecraft, possibly igniting the Agena's 4,000 pounds of fuel. That's just what he needed: a fireball above China. Mao Tse-tung, the chairman of China's communist party, would go berserk. Neil would have to use *Gemini 8*'s maneuvering thrusters to undock, to back away from the Agena, but that could rip the whole damn thing apart and shower their orbit with little pieces and . . . And God could let the Sun fall from the sky! Instantly it was clear. He had one choice. If Agena's fuel didn't blow, they would soon pass out from the centrifugal force of the increasing spin and if they were going to die anyway, he reasoned, screw it. Let's take our best shot.

"We're going to disengage," Neil told Dave who immediately agreed, and suddenly Neil's hands were magic. He was firing bursts from the Gemini's maneuvering thrusters and he fought and fought until Scott was convinced Neil had steadied the linked spacecraft enough to undock and Dave yelled, "Go," and he hit the undocking button as Neil gave the thrusters a long hard burst.

"Come on," Armstrong shouted, and there immediately came a bang from the motors driving the docking latches open and the docked ships freed themselves with *Gemini 8* pulling straight back. It was like two people facing one another and each using their hands to push the other away. For every action there's an opposite reaction, Neil knew, which dampened his concern over the distance between them and the Agena. With no resistance in space, *Gemini 8* and its target rocket should continue moving apart in their separate orbits and . . .

Neil fires his thrusters to disengage from Agena. (NASA)

What's this? *Gemini 8* was beginning to spin faster. Instantly the astronauts realized the problem was not in the Agena, it was in *Gemini 8*. It was now increasing its spin, making them even more dizzy, and Neil Armstrong's analytical mind stepped up to the plate. Despite the growing chaos, he was quick to realize his spacecraft's spinning was a disaster in the making. *Gemini 8* was an out-of-control dilemma that would keep growing until the centrifugal forces ripped everything apart and he and Dave would not only lose their focus, they would lose their vision and their consciousness.

To stabilize his spacecraft *Gemini 8*'s commander recognized he needed an independent control system. One stronger than his attitude-control thrusters, and that could only be *Gemini 8*'s RCS (Reentry Control System).

The RCS had two individual rings of hypergolic rockets. They were more powerful than the ship's troublesome attitude-control thrusters, but they were a separate part of the spacecraft for one reason. A fresh, independent control system was a must to position *Gemini 8* correctly for its return to Earth. Without it, Neil and Dave would simply not get home. They would burn up during an out-of-control reentry.

The book says that once you activate your two RCS rings, A and B, you must then land at the first opportunity.

But Neil also knew they had only one real hope. Use *Gemini 8*'s primary reentry control rockets.

He was thinking again that he and Dave were going to die anyway so what the hell, with their chances being slim to none, he'd take slim every time.

Both astronauts wavered on the edge of losing consciousness and when they first tried to fire their reentry thrusters nothing happened. After turning the system off, then trying again, the thrusters responded, and Neil was busy trying to stabilize *Gemini 8* when they came in contact with the tracking ship *Coastal Sentry Quebec*.

James Fucci, the CapCom aboard, was fussing because he couldn't get a solid signal, unaware that the spacecraft was rolling.

"*Gemini 8*, CSQ CapCom," Fucci called. "Com check. How do you read?"

"We have serious problems here," Dave told him. "We're spinning and we're disengaged from the Agena."

"Okay, we've got a 'Spacecraft Free' indication here," Fucci told the crew. "What seems to be the problem?"

"We're rolling," Armstrong told him.

Houston Mission Control was listening and suddenly the room was fully awake and alert.

Rookie CapCom Bill Anders managed one word, "Roger."

"*Gemini 8*, CSQ."

"Stand by," Neil told the tracking ship.

"Roger."

Those listening on the worldwide network were not aware of just how dangerous the situation was aboard *Gemini 8*, and they stood by while Neil fought to save his spacecraft, firing his reentry thrusters, doing everything he could to regain control with Dave quickly responding to everything he asked.

They fought and the ground stood by and after 37 seconds of silence, Dave told them, "Okay, we're regaining control of the spacecraft slowly, in RCS direct." "Roger, Copy," Bill Anders acknowledged, and thumbs went up in Mission Control.

"We're pulsing the RCS pretty slowly here so we don't control roll right," Neil told the ground. "We're trying to kill our roll rate."

"Okay, fine," Anders acknowledged. "Keep at it."

"8, CSQ," James Fucci called from the *Coastal Sentry Quebec*. "How much RCS have you used and are you just on one ring?"

"That's right," Neil quickly confirmed, knowing what the tracking ship was getting at. "We are on one ring, trying to save the other ring for reentry."

CapCom Bill Anders caught a hint of a "can do" in Neil's voice, and he told the astronaut, "Everything's okay."

And it was.

Once the firing of the reentry thrusters had stabilized *Gemini 8*, Neil and Dave began firing and rechecking their thrusters one at a time. When they hit the switch for number eight, their spacecraft began to roll again. They had found the culprit.

Gemini 8's emergency had all the lights on at Houston's Manned Spacecraft Center with the lights in Mission Control burning the brightest. Every single person there was hanging onto every single word from the Gemini control commentator:

We are 8 hours and 3 minutes into the flight of *Gemini 8*. And in view of the trouble encountered at 7 hours into the flight as reported earlier, the flight director has determined to terminate the flight in the 7-3 area. We plan to bring the flight down on the 7th orbit in the 3rd, what we call the 3rd zone, which is approximately 500 miles east of Okinawa, it's in the far west Pacific. Our situation out there is as follows—a destroyer named the *Mason* is about 160 miles away at this time, it is proceeding towards the point, which should come very close to the . . . well it may be a little delayed, get there after the landing itself. The first estimate I have on the retrofire time is ten hours and four minutes into the mission, in other words two hours from now. Landing should take place some 25 to 26 minutes later.

In addition to the *Mason*, a C-54 has been dispatched from Tachikawa Air Force Base in Japan; it's proceeding to the point. Another C-54 is proceeding to the point from Okinawa. Another location here on the landing point is quoted to me as 630 nautical miles south of Yokosuka, Japan. The weather conditions out there are partly cloudy, visibility

10 miles, and the landing will be made in full daylight. It's 12:30 P.M. out in the 7-3 area.

This is Gemini Control at 8 hours, 6 minutes into the flight.

Riding backward with only a disappearing band of blue on the horizon, Neil and Dave were ready to begin their emergency reentry in total darkness.

High over the Arabian Sea Neil Armstrong hit *Gemini 8*'s brakes. He fired their four retro-rockets beginning a never-before-flown reentry through the dark

During the mission following his emergency return from orbit, Neil Armstrong listens intently in Mission Control to the troubles his friend Gene Cernan was having trying to maneuver in *Gemini 9*'s EVA without support devices. (NASA)

night over India, Burma, and then into sunrise over China and the East China Sea. All the way through reentry, they could not expect contact with any tracking station or ship if needed. The two astronauts were it. They only had themselves.

In Neil's home Janet Armstrong stared at her NASA squawk box. The speaker relayed the astronauts' transmissions along with the mission commentator's explanations. By her side was world-renowned *Life* magazine photographer Ralph Morse. *Life* magazine had a contract with the astronauts to tell their story—a clever way for the space flyers to raise their pay above the military grade, and their growing families were thankful.

Ralph Morse's photographs from the world's hottest assignments weren't only renowned; Morse enjoyed a reputation as a very likeable fellow. He was a friend to most and Janet found comfort with Ralph in her home. She needed moral support during Neil and Dave's life-threatening plunge through the pitch-black atmosphere.

Janet tried desperately not to miss a single word from Gemini control:

The pilots were counted down in the blind via the Kano, Nigeria Station, and Neil Armstrong, while he said nothing leading up to the point of retrofire, came back with a very reassuring "We have all four retros, all four have fired." A cheer went up here in the control center, and I'm sure everyone can understand why. The contact from here is through the Kano station. Armstrong, after the maneuver, went on to relate that everything was in a stable condition and seemed to be proceeding satisfactorily. We want to emphasize again that there is practically no communication expected now for some time. We are going to try to reach them via the *Coastal Sentry Quebec* on HF after they emerge from blackout, but that signal at best will be marginal. Probably our first authoritative information will come via one of those C-54 aircraft maneuvering in the area east of Okinawa.

"This is Gemini control, Houston."

Neil and Dave shifted in their seats for comfort as their spacecraft edged into the atmosphere. Heat built up. *Gemini 8* swayed slightly in its automated and meticulously computer-controlled reentry. Any departure from its established protocols could have serious repercussions and then Neil would have to take over manually and fly the danger-filled reentry.

"I keep thinking there's something we've forgotten," he told Dave. "But I don't know what it is."

"As far as I know, we've done everything," Dave reassured him.

Suddenly the two flyers were enveloped by what appeared to be a devouring fireball in their dive earthward. All they could do was monitor their spacecraft's systems as it carved an ionized tunnel through the thickening air.

In this ionization envelope there were no radio signals in and none out, and Neil and Dave were riding little more than a blazing meteor. But despite their apprehension, they were cool inside as they watched the brilliant orange teardrop grow—its flames burning their path through the atmosphere.

All the while Janet and Ralph were straining to hear any report through the broadcast static and those in Mission Control were feeling equally helpless. There was simply no way to reach *Gemini 8*, no way to talk to Neil and Dave. Those on the ground were in a sense as alone as the two astronauts.

Everyone waited and the *Gemini 8* astronauts felt their weight grow as they entered the full deceleration of their reentry, and with the slowing spacecraft came a more gentle oscillating from side to side and suddenly Neil and Dave were at 50,000 feet. The reentry from hell was almost over, and they were enjoying a bright day with their normal weight as their drogue chute came out, stabilizing their spacecraft for its main parachute to blossom and drop them safely into the Pacific 480 miles east of Okinawa.

A recovery aircraft parachuted in three rescue swimmers, who took their positions alongside the floating *Gemini 8* and Mission Control reported: "The swimmers in the water have been in voice contact with the crew. The astronauts report all on board is well and they say their condition is okay. Neil and Dave are standing in their seats waving and smiling."

Janet Armstrong jumped to her feet and began clapping and laughing. "Thank goodness," she said with a big smile. "Thank goodness." Being the wife of a test pilot wasn't the easiest job around.

Neil and Dave had performed their reentry with great skill. They had to end their flight early with Neil telling the media, "We're disappointed we couldn't complete the mission, but the part we did complete, and what we did experience, we wouldn't trade for anything."

Gemini 8 safely back on Earth—on the Pacific Ocean anyway. Left to right: Dave Scott and Neil Armstrong sit in their spacecraft secured by a flotation collar supplied by three Navy Rescue swimmers. (NASA)

For weeks to come the crew of *Gemini 8* was subjected to a number of debriefings and reviews with bosses and Gemini support groups—each group with a special interest.

Time and time again Neil and Dave told them everything they knew and the second-guessing began. Why didn't Neil do this? Why didn't he do that?

John Glenn called the criticism of Armstrong's performance aboard *Gemini 8* nonsense. "You'll never hear it from me," Glenn said emphatically. "I don't think anybody was as experienced a pilot as Neil was at the time. He assessed when it was getting beyond his control, and he assessed it right."

Joining Glenn in Armstrong's defense were two of the top managers in

NASA, Chief Flight Director Chris Kraft and the director of the Manned Spacecraft Center Dr. Robert Gilruth.

"Armstrong's touch was as fine as any astronaut," said Kraft, a veteran flight director. He quickly added, "Neil calmly reported the emergency, and when we learned the crew was being tossed around and beginning to suffer from greyed-out vision, it was clearly a life-threatening situation in space. The worst we'd ever encountered.

"Dr. Gilruth and I both first thought Neil was having trouble with the stick. It never occurred to us that he had a stuck thruster. If we had heard about the problem when they were still docked, we would have told them to do exactly what they did, 'Get off that thing!'

"The spin rate was up as high as 550 degrees per second." Kraft continued. "That's about the rate at which you begin to lose consciousness or the capability to operate. Neil Armstrong realized they were in very serious trouble, and he took all the power off the Gemini to try to stop the spin, and then figured the only way to recover was to activate the reentry attitude-control system. That was truly a fantastic recovery by a human being under such circumstances and really proved why we have test pilots in those ships. Had it not been for Neil's good flying, we probably would have lost that crew."

Most NASA managers in the agency top tier like Kraft and Gilruth were pleased with Neil and Dave's performances, but for Neil and Dave their first concern was What did Deke think? Deke Slayton was their immediate boss and they knew their future in spaceflight pretty much rode in his hands.

Once Deke settled the facts of the *Gemini 8* emergency in his mind, he came to the conclusion that Neil Armstrong's abilities to reason, to think, to handle emergencies, to fly the hell out of anything from the Wright brothers planes to rocket ships, and the great support provided by Dave Scott, made them both leading candidates to land Apollo's lunar modules on the moon.

Neil would command *Apollo 11* and Dave would command *Apollo 15*, a fact that Deke would keep to himself until the right time, and he moved them back into the flow of things, telling Neil and Dave to go and help those trying to solve spaceflight's EVA problems.

Dave Scott remained disappointed he had not been able to try out his planned two-hour spacewalk. He had hoped to solve some of EVA's difficulties, which had first appeared when Ed White had so much trouble getting back in *Gemini 4*.

He and Neil joined the Gemini flow and a year after Ed White made the first spacewalk, astronaut Gene Cernan was ready. With Dave's disappointment, Cernan would now be the second American to walk in space.

Two-and-a-half-months after *Gemini 8*, *Gemini 9*'s Tom Stafford and Gene Cernan reached orbit on June 3, 1966, and Cernan began his spacewalk attached to a 25-foot lifeline. He looked forward to experiencing the freedom that Ed White had on his spacewalk. But Cernan didn't find it. White had gone outside just to skip and flit about for a few minutes. Cernan had gone out to work. He had specific duties, but quickly discovered he was little more than a clumsy sloth climbing a greased pole.

Mission planners had installed in the back of *Gemini 9*'s equipment bay a backpack called an AMU (Astronaut Maneuvering Unit). Attached to a long 125-foot tether, the backpack's jets would enable Cernan to maneuver alongside *Gemini 9*.

But the spacewalker soon learned he couldn't actually "fly" in space. Once away from his mother ship, he found himself in an independent orbit, and changing it was a major chore. Cernan wasn't having a glorious experience. He could only move about an inch at a time. He did everything he could to clamp his gloved hands on something that would secure him to his ship. But without proper handholds, he kept slipping off the sleek hull of *Gemini 9* and having to fight his way back. He needed a jet gun like the one that Ed White had used. Cernan could damn well use his backpack if he could get to it, but the 15-foot walk to *Gemini 9*'s rear equipment bay was proving to be impossible. He had thought it would only take a few minutes and now he'd been trying to get there for nearly an hour.

He was clumsy, frustrated, and struggling fully exhausted, but the astronaut who would be the last to walk on the moon, finally made it. "It's a strange world out here!" he radioed Tom Stafford in *Gemini 9*'s commander seat, barely adding, "Whew!"

"Take a rest," Stafford told him.

He didn't have to be told twice. Cernan was grateful just to hang onto some equipment in the rear of his ship. After a brief rest he began trying to maneuver his pressurized, suited body into the backpack so he could possibly jet around. This would be real maneuvering in an EVA, but in spite of all his efforts, trouble remained his companion.

The job of slipping into his backpack demanded more than slipping into simple straps. He needed to make electrical and other connections, and Cernan found every move was more time-consuming than he had planned. When he seemed to have one of his backpack hookup procedures under control, he floated from the spacecraft. He had nowhere to stand—nothing to hold him in place. There was no way he could maintain a solid plant of his body. There were a few footholds and handholds, but they were woefully inadequate. He needed positions that would allow him to use leverage. Soon he was severely overworking his own chest pack, which circulated oxygen through his suit and also removed excessive moisture from his body. He perspired. Fog collected inside his helmet visor and froze, and he endured excessive heat, perspiration, and ice all at the same time. He was barely able to see through the visor—a potentially lethal situation for a man turning like a bloated rag doll in a vacuum, only several feet from the security of his spaceship's cabin.

"We've got problems," Tom Stafford told Mission Control. "Gene is fogging up real bad." Then he spoke only to Cernan on *Gemini 9*'s crew loop. "How bad is it, buddy?"

"I'm really fogged up, Tom," Gene told him. Stafford didn't like the sound of Gene's voice.

Stafford called Mission Control again. "The pilot's visor is fogged over, and I'm having trouble understanding him. He sounds like a large gargle. If the situation doesn't improve . . ." *Gemini 9*'s commander suddenly went quiet and came back to Mission Control with his own command decision. "It's no go on the AMU! The pilot is fogged up!"

"We confirm, no go," Mission Control agreed, and now Gene Cernan's problem was figuring out how to get the hell back inside. He began trying to return to his seat in *Gemini 9*, carefully, hand over hand, slipping, fighting every inch of the way. He was only able to see through small sections of his visor where his breath had melted the frost.

"He's in trouble," said one flight surgeon in Mission Control. Gene Cernan's heart was beating 180 beats per minute. Suddenly Tom Stafford felt icy chills as he recalled a very private conversation he had had with Deke Slayton before launch. "Look, Tom, what Gene's going to do out there is pretty risky. If he's outside and he's in trouble . . ." Deke paused. "If for some reason the spacecraft is in trouble, short of fuel, or something, and well, if there's no chance of getting him back in . . ."

Deke's voice faltered, then the words returned, "There's nothing written on this. No mission rule, but, well, I think you understand. If that kind of thing happens, you've got to cut him loose."

No one on the ground was more concerned for Gene Cernan than Janet and Neil Armstrong. Janet and Gene were classmates at Purdue. Tom Stafford would never learn as commander what he would have done if faced with having to cast his buddy and fellow pilot to certain death.

He was so grateful he didn't have to make that decision. He was never so damn relieved as he was to see Gene Cernan fighting and clawing and struggling through his hatch, pulling his exhausted body back into his *Gemini 9* seat.

When they finally had their spaceship all buttoned up, and they were safe, the third man to walk in space, the second American, had been outside a record two hours and nine minutes, and all of it had been pure hell.

Following *Gemini 9*, both Mike Collins on the *Gemini 10* and Dick Gordon on the *Gemini 11* experienced troubles maneuvering away from their spacecraft. Collins, who used a jet gun to move to a nearby Agena target rocket, told flight planners, "I found that the lack of a handhold is a big impediment. I could hang on to the Agena, but I could not get around to the other side where I wanted to go. That is indeed a problem." Gordon, like Cernan, became hot and sweaty and his visor fogged. "I'm pooped," he said, deciding to cut short his walk.

To date the Gemini program had been highly successful. It was spaceflight at its best, developing systems needed to reach the moon. Only EVA was a problem. Astronauts' helplessness in moving about outside their spaceships left NASA's bosses in a quandary. "How can we send astronauts to the moon, if . . . ?"

Deke Slayton, Neil Armstrong, Tom Stafford, and Gene Cernan—all those working on EVA knew most definitely they were left with one shot. They were down to one flight, *Gemini 12*.

That was the bad news.

The good news was veteran astronaut Jim Lovell would command *Gemini 12* and Buzz Aldrin, who had one of the astronaut corps' most educated scientific minds, would attempt Gemini's final EVA.

Aldrin graduated third in his class from West Point with a degree in mechanical engineering and he received a Ph.D. in the science of astronautics from MIT. Aldrin was not only smart and educated, he was a tinkerer. He built on the experience of the others. He talked at length to Cernan, Collins, and Gordon and he listened. He asked them questions again and again and took their advice. He gathered special devices like a wrist tether and a second tether much like those used by window washers. For foot restraints, he fashioned himself a pair of "golden slippers." He would bolt them to the workstation floor he had put in his spacecraft's equipment bay. There he would bring each major job to be done as if he was bringing work home to his garage workshop.

When it came to moving easily from point A to point B, Buzz fashioned portable handholds he could slap onto either the hull of Gemini or the hull of the Agena to keep his body under control. And to make sure he never found himself without a needed tool, he placed a variety in pockets and holding places within his spacesuit.

Aldrin nodded ready to Lovell and on November 11, 1966, *Gemini 12* roared into orbit. It captured its Agena target rocket, and once docked, Buzz Aldrin, who would make the first landing on the moon with Neil Armstrong, once and for all banished the gremlins of EVA.

First Buzz took what he had learned and recognized movement in orbit is strange to our senses. A circular orbit is like trying to walk on a merry-go-round platform. Straight is not what it seems. If you move perpendicular to the line from the satellite to Earth's center, the forces are small, and the maneuvers are easy. If you move any distance directly away from, or toward Earth's surface, the forces can be noticeable, although moving inward should be okay, but moving out is not. The physics is simple, and Aldrin proved such a master at it that he seemed more to be taking a leisurely stroll through space than attacking the problems that had frustrated and endangered previous astronauts. He erased NASA's grave doubts about its chances of landing astronauts on the lunar surface.

Aldrin moved down the nose of the Gemini to the Agena like a weightless swimmer, working his way almost effortlessly along a six-foot rail he had locked into place once he was outside. Next came looping the end of a 100-foot line from the Agena to the Gemini for a later experiment, the job that had left Dick Gordon in a sweatbox of exhaustion. Buzz didn't show even a hint of heavy breathing, perspiration, or an increased heart rate. When he spoke, his voice was crisp, sharp, clear. What he did seemed incredibly easy, but it was the di-

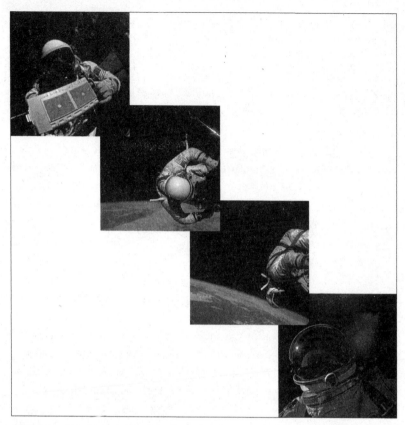

Using his self-fashioned EVA aids, Buzz Aldrin works effortlessly outside *Gemini 12*. (NASA)

rect result of his incisive study of the problems and the equipment he'd brought from Earth. He also made sure to move in carefully timed periods, resting between major tasks, and keeping his physical exertion to a minimum. When he reached the workstation in the rear of his ship, he slipped his feet into his "golden slippers" and secured his body to the vehicle with the waist tether.

Buzz hooked different equipment to *Gemini 12*, dismounted other equipment, shifted them about, and reattached them. He used a unique "space wrench" to loosen and tighten bolts with effortless skill. He snipped wires, reconnected wires, and connected a series of tubes.

Mission Control hung on every word exchanged between Aldrin and Lovell and asked, "Buzz, how do those slippers work?"

"They're great, just great," Buzz smiled. "I don't have any trouble positioning my body at all."

Neil Armstrong was in disbelief listening to Aldrin lay the blueprint for a successful EVA. Neil said quietly, "Buzz you can fly with me anytime," and would he ever.

At the end, Aldrin's doctorate in astronautics from MIT solved the final critical procedure moon-bound astronauts would need to make it all the way to their destination.

In ten flights and twenty months Project Gemini had come from behind in America's fierce competition with the Russians.

Cosmonauts were still clumsily plodding through orbit compared with the freewheeling antics of the Gemini crews.

No cosmonaut had yet mastered rendezvous. No Russian spacecraft had yet docked with one another. They had conducted only one EVA, and were far from the EVA skills Buzz Aldrin had just given the science.

Apollo was next. America was on a roll, moving fast to fulfill President Kennedy's promise to send a human to the moon before the 1960s ended.

But in only days beyond the holiday season, NASA would learn it was moving too fast. America would pay a terrible, heartbreaking price to be first on the moon.

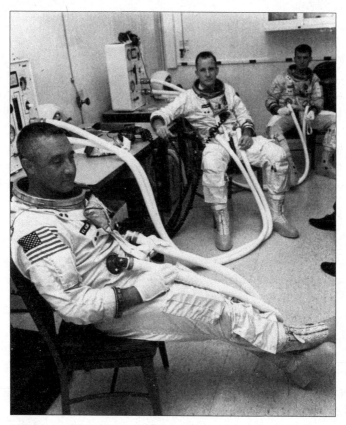

Gus Grissom, Ed White, and Roger Chaffee suited up for their *Apollo 1* launchpad test. (NASA)

TWELVE

TRAGEDIES GROUND SPACEFLIGHT

With the successful conclusion of Project Gemini, NASA's future appeared bright but that wasn't altogether so. Grumblings in the halls of Apollo could be heard as astronauts Gus Grissom, Ed White, and Roger

Chaffee were boarding on its launchpad the first problem-filled copy of Apollo. It was to be an important but hurried ground test.

At the same time five astronauts including Neil were reporting to the White House for glad-handing duties. President Lyndon Johnson had invited Neil, Gordon Cooper, Scott Carpenter, Jim Lovell, and Dick Gordon to the country's first home to witness the signing of an international treaty, one with too long and too complicated a name for anyone to remember. It was simply called the "non-staking-a-claim treaty."

The agreement was also being signed simultaneously in other world capitals including London and Moscow. It stated simply that those signing agreed not to claim any land on the moon, Mars, or any other heavenly body. It also guaranteed the safe and cordial return of any Homo sapiens flying through space who landed unexpectedly within the borders of another country—a clause Neil greeted warmly, remembering *Gemini 8*'s landing trajectory could have brought him and Dave Scott down in China.

Following all the jawing and official doings, Neil and the other astronauts attended a reception in the White House Green Room.

Each astronaut held a glass of wine—or something a bit stronger—and smiled with proven pleasantries. They rubbed elbows with many of Washington's movers and shakers including foreign dignitaries such as Russia's Ambassador Anatoly Dobrynin. Then Neil watched the Russian diplomat hurriedly leave the reception after being approached by a whispering aide.

The same types of hushed conversations were spreading around the White House Green Room—especially among NASA's top brass. Neil gave little notice, pleased to see the reception end precisely at 6:45 P.M. as promised.

Scott Carpenter left for the airport while Neil and the other astronauts took taxis to the Georgetown Inn. They arrived about 7:15 P.M. going directly to their rooms where they found their phone's red message light flashing.

Each had an urgent call. The message gave them a number to phone. The person answering in the Apollo Program Office told Neil there had been a fire on pad 34 and most likely the *Apollo 1* crew hadn't survived. The details were still sketchy and he and the other astronauts shouldn't talk to the media or anyone. The person strongly suggested Neil and the others disappear until further word.

Neil stared at the phone, and then hung it up. He had heard all the stories about the first Apollo being a shoddy piece of crap and that Gus had hung a lemon on its simulator. He quickly phoned a friend involved in the test.

The friend told him apparently somewhere beneath the seat of Gus Grissom there were unbundled live wires that had been walked on by technicians again and again and one had apparently been chafed. Its insulation was worn and torn. The wire lay bare in a thick soup of 100 percent oxygen and the wire's flammable materials exposed to an ignition source were extremely dangerous.

Pure oxygen had been used in the Mercury and Gemini without trouble. But the amount of pure oxygen filling a ship as large as Apollo was another story and Neil shook his head in disbelief.

Gus Grissom had apparently shifted his body for comfort.

His seat moved the bare wire and a surge in voltage followed.

The wire must have sparked.

Flames filled *Apollo 1*, feeding on the rich oxygen-soaked flammable materials surrounding the astronauts.

The crew needed one-and-a-half to two minutes to open *Apollo 1*'s hatch but the fire took the crewmen's lives in a matter of seconds.

"Damn," Neil cursed. "To die in a ground test?" he shouted aloud. "It's bad enough to die in flight, but to die in a ground test?" Neil didn't hide his disgust. "Hell, that's an indictment of ourselves. That sort of thing shouldn't happen on the ground," he insisted. "We should get it right before we expose ourselves to any such risks."

He was mad at everyone yet at no one, and picked up the phone again. He tried to call his wife. There was no answer.

What Neil didn't know was that astronaut Alan Bean had already phoned Janet and asked her to go next door and stay with Pat White. "There's been an accident at the Cape," Al Bean told her. "We don't know how bad yet, just stay with Pat if you can, Janet, until we know."

Janet had walked over to the White's to find Pat not there. She had gone to pick up her and Ed's daughter and son and Janet waited. Only minutes passed and Pat, with Bonnie and Eddie, drove into the carport. Janet told her as calmly as she could there'd been a problem and she helped Pat bring in groceries she had just bought.

Meanwhile down the street neighbor astronaut Bill Anders was being sent by Deke to tell Pat there had been a fire on the pad and Ed, Gus, and Roger didn't make it. They had been killed.

Anders didn't want the job, but someone had to do it, and he rang the doorbell. He told Pat as delicately as he could, and she instantly became distraught collapsing in his arms. Janet consoled her and the children and the neighborhood rallied with sympathy and full support.

Other astronauts at the homes of Gus Grissom and Roger Chaffee repeated the scene. It was a scene test pilots' families knew only too well. It was part of the life they had freely chosen and Neil, back in his hotel room, was feeling the pain in his own private way.

He stood before his hotel window looking out on the night's lights illuminating official Washington. The country's institutions and monuments were all lit but at the moment they commanded little of his attention. His thoughts and pain were back in the neighborhood, back with family. He wished badly to be with everyone remembering Ed and Gus and Roger.

Neil had known Gus Grissom for a long time. They'd been friends as far back as Edwards, and he had grown close to Ed White with him being one of the best neighbors and friends he'd ever had. They had hunted, fished, flown together, owned their own plane together, and the night Neil's house burned, how could he forget Ed White was the first there to help fight the flames and save Janet and his boys.

Neil warmly remembered Ed taking on that big house fire with only a garden hose—in his bare feet no less. And when it came time to rebuild, Ed was there again. Stacking lumber, pushing a wheelbarrow, even pounding nails with a hammer. He remembered Ed pushing the wheelbarrow full of concrete and yelling, "Put that wheelbarrow down, White. You Air Force guys don't know anything about machinery," he joked.

And there were other laughs—not all in his favor—such as the time in astronaut water-rescue training when Neil had gotten himself in over his head. Guess who saved Navy? Air Force, that's who, the all-American athlete from West Point.

Neil only wanted to remember the good about Ed, Gus, and Roger, but he couldn't push aside the bad—the way they died. It hurt too damn much and he walked away from the window and headed to the door. He needed to get out. For once Neil Armstrong needed company and he headed into the hallway to find out what the other astronauts had heard.

Within hours the *Apollo 1* crew was being remembered in America's homes. In the home of Frank Sinatra the memories were fresh. Ten days before flying to the Apollo plant in California for simulator training they ran into problems with one of their T-38 jets. The official logs for the astronauts' T-38s indicated an abundance of problems when the aircraft neared Las Vegas, and often they had to land at nearby Nellis Air Force Base for maintenance. Naturally, while their jets were being checked and serviced, the astronauts had to take in a show on the Vegas Strip.

On that particular day, Frank Sinatra was performing, and no sooner had Gus, Ed, and Roger sat down, Frank had them brought up front. They were wearing their astronaut flight jackets and Ol' Blue Eyes took a shine to Grissom's jacket. He was especially impressed with its mission patches.

Gus grinned. He stood up, removed his jacket, and gave it to Frank. Sinatra was so moved he cried right there on stage.

Tonight, ten days later, Frank Sinatra was crying even more.

NASA buried its dead—Gus and Roger at Arlington, Ed at West Point—and turned its attention to the red flags raised by the *Apollo 1* fire.

NASA Administrator James Webb got the message: *Fix Apollo!*

John Glenn escorts Gus Grissom to his Arlington grave. (NASA)

He gave the job to Floyd L. Thompson, director of the Langley Research Center in Virginia, who set up a board of review with members from the best sciences and best investigators in the country.

"Nail the problem," Thompson told them. "We must get to the moon."

Neil and the other astronauts in line to command an Apollo flight—Wally Schirra, Frank Borman, Jim McDivitt, Tom Stafford, Pete Conrad, Jim Lovell, Alan Shepard, and Gene Cernan studied thousands of dials, switches, transistors, and electrical connections while ripping everything that would burn from the Apollo spacecraft on the construction line. They replaced it all with fire-retardant materials. If it burned it didn't go into the new Apollo, and they added a dual life-support system, which re-created Earth's atmosphere. Inside the new spacecraft the crew could breathe normally and fire would burn normally. And they installed the best available fire-extinguishing systems and a quick-opening hatch. If they, or any other astronauts, had to fight fire on or while traveling to and from the moon, they would at least be dealing with fire under familiar conditions.

To verify their findings and thinking, the astronauts, board members, and investigators built an exact copy of *Apollo 1* and set it ablaze. The test shook them so badly some went home and stared at the walls. They badly wanted the heads of those responsible for the deaths of Gus Grissom, Ed White, and Roger Chaffee. Their search wasn't at all difficult. The incompetence, inexperience, and laxity were right in front of their eyes, and the astronauts could swallow no more.

"We need a no-nonsense ramrod running Apollo," Neil Armstrong said flatly. "Gus, Ed, and Roger's deaths have given us the gift of time. We didn't want the gift, but by god we've got it. We have months to not only fix the spacecraft," he added, "we must rethink all our previous decisions, plans, and strategies, and change a lot of things for the better."

Neil and the astronauts were throwing down the gauntlet, and NASA's George Page, a veteran of the Mercury and Gemini programs, picked up the challenge.

"I have your man," Page told them. "He's the guy that cleaned up the Atlas mess. He's the one that launched John Glenn, Scott Carpenter, Wally Schirra, and Gordo Cooper in orbit."

"What's his name?" Neil asked.

"T. J. O'Malley," George Page smiled. "He can make the trains run on time."

Wally Schirra and Alan Shepard's faces lit up. "We know the altar boy," Wally said. "Get 'im," echoed Alan.

Page reached for his phone. T. J. was in Quincy, Massachusetts, working for General Dynamics' electric boat division, and the phone only rang once.*

---

* T. J. O'Malley was my poker buddy who nitpicked my writing severely. I made a practice of making sure what I wrote about him was precise. Believe me, the following conversations in this chapter between O'Malley and his wife, Ann, as well as with his friend, George Page, are verbatim.

"O'Malley."

"T. J., this is George."

"Hey there, Mister Page." O'Malley's demeanor changed instantly. "How's everything at the Cape?"

"A mess," George Page said flatly. "No, let's make that a goddamn mess."

"How many asses have you hanged for that fire?" O'Malley grunted.

"Not enough," Page said. "If we hanged all we should, we'd run out of rope."

"That bad, huh?"

"That bad, T. J." He paused before pleading, "We need you, old friend. We need you to come back and take charge of the Apollo spacecraft."

O'Malley grunted again.

"How about it?"

"Well, George." T. J. turned serious. "I'm expecting a promotion here. In fact, I'm expecting it tomorrow, and—"

"T. J., we're in a terrible mess," Page interrupted. "We need you," he pleaded again. "The astronauts want you, the country needs you. If we're going to make it to the moon—"

"I'll think it over tonight," T. J. interrupted this time. "I'll get back to you tomorrow morning, George. I promise."

The two close friends hung up and George Page spent the night with his fingers crossed.

That evening Thomas J. O'Malley turned to Ann and said, "Mrs. O'Malley, George says they need me at the Cape. He wants me to come down and see what I can do with that mess left by the launchpad fire."

"Tom," she said lovingly, "no doubt they need you, and no doubt you'd be the man for the job, and you've always taken care of your family, and we could do no less than support you, and honey," she winked, pointing at the snowbank on their lawn, "it's your decision but it sure is nice this time of year in Cocoa Beach."

Thomas J. O'Malley returned to the rocket pads of Cape Canaveral exactly one year to the day after he had left, and was immediately hired as the vice president of Apollo operations by North American Aviation. He was given the job to get America to the moon before the 1960s were over. Experience and know-how suddenly counted, and T. J. O'Malley went to work that afternoon. By sunrise the next day, Tom O'Malley knew two things: George Page was right, and he wasn't at all sure Page was his friend.

The Apollo team was overloaded with retired military—colonels and generals—doing little more than flying on North American's travel vouchers to military reunions and such. Each was buying his stuff—each only accountable to self. And experience? Hell, they had none of that.

The whole thing smelled of military paybacks for aviation and aerospace contracts, and in the coming months while the review board investigated, while others were pointing fingers and protecting their own backsides, T. J. O'Malley put on his boots and began kicking ass and taking names.

North American's most qualified and experienced aerospace engineers got busy redesigning and rebuilding Apollo to the specifications of the astronauts, and Neil Armstrong got busy learning how to fly one of the redesigned Apollo spacecraft. He knew Gus, Ed, and Roger's deaths had given them the gift of time needed and by god, if Neil A. Armstrong was to have a say, they'd get it right for three fallen friends.

He also knew the crew selected to fly to the moon would be going on the backs of the *Apollo 1* astronauts and he managed a smile. "Thanks guys," he spoke quietly. "We'll get 'er done."

Neil and the other astronauts hadn't a clue that Russia's cosmonauts were about to run into their own tragedy dogging their efforts to also reach the moon.

Less than three months after the *Apollo 1* fire, Cosmonaut Vladimir Komarov rode the first Soyuz spacecraft into Earth orbit. He had one comment: "Splendid."

The Soviet Union's senior cosmonaut judged their all-new spacecraft as promising and following a day of testing, the plan was for a second Soyuz to join him in orbit. The two spaceships would then dock and their two spacewalking pilots would switch spacecraft. The feat would put Russia's cosmonauts back on par with America's astronauts—a capability Russia needed to reach the moon. But no sooner had Komarov settled in while racing around Earth than the all-new Soyuz developed serious problems.

Unlike America's Gemini and its future Apollo that ran on battery and fuel cell power, Soyuz was built to draw its energy from two solar panels.

The right panel extended.

The left panel did not. It remained closed. Nothing Komarov tried would release the solar wing.

By the second orbit, Russia's mission control, located a few miles outside of

Moscow, was on full alert. *Soyuz 1* was receiving barely half the electrical energy needed. Its systems and controls were failing.

Komarov's shortwave radio transmitter died. His ultra-shortwave radio remained operational, but barely. He could only communicate with mission control sparingly and he received instructions to change the attitude of Soyuz's face to the sun. The move could help pull in more power. It just might give him enough juice to free the jammed panel. It didn't.

By the fifth orbit the new Soyuz began shredding itself and mission control feared for Komarov's survival. His power was failing. Communications began to break up. He shut down the automatic stabilization system and went to manual control with his attitude thrusters.

The thrusters operated only in balky spurts. Komarov felt the ship getting away.

Controllers instructed the cosmonaut to let Soyuz go, let it drift. He would soon be moving through a series of orbits during which he would not be able to maintain voice contact with his flight controllers. Between the seventh and thirteenth orbits he would be away from tracking stations, out of touch.

This was the so-called daily "dead period" for Russian spacecraft and mission control instructed him to try to get some sleep. Then he moved out of radio range. He would be in a communications blackout for the next nine hours.

The time passed slowly. At the close of the thirteenth orbit they heard Komarov's voice. His report sent chills through the control center. The ship was dead. Manual control with sputtering thrusters was sporadic at best. Soyuz was rapidly becoming a careening, wobbling killer with its pilot trapped inside.

Officials canceled the launch of the second Soyuz. They told Komarov they had to gamble to get him back to Earth. He would have to fire his retro-rockets for reentry on the seventeenth orbit and use all his strength and knowledge to try to manually hold Soyuz on a steady course. He would have to maintain attitude control through the fireball of reentry.

They didn't have to tell Vladimir Komarov why.

The flight director said aloud what Komarov and everyone knew. "He is out of control. The spacecraft is going into tumbles that the pilot will have difficulty stopping. We must face the truth. He might not survive reentry."

As a select few in NASA listened to a translation from America's intelligence-gathering assets Neil Armstrong immediately recognized the cosmonaut's problems. He and Dave Scott had faced much the same on *Gemini 8*.

In early 1958, a few months after the Soviet Union launched Sputnik, President Dwight Eisenhower authorized the development of a top-priority reconnaissance satellite project operated by the CIA Directorate of Science and Technology with assistance from the U.S. Air Force. It was used to "eavesdrop," and for photographic surveillance of the Soviet Union, China, and other areas from 1960 until May 1972. The project's code name was Corona, but to hide its true purpose, it was given the cover name Discoverer and said to be a scientific research and technology development program.

In Russia's mission control the flight director picked up his telephone and issued orders. A powerful car pulled up before a Russian apartment complex. Two men rushed into the building, emerging moments later with a woman. The car roared off toward the control center.

The woman was Valentina Komarov, the cosmonaut's wife and the mother of their two children.

By the time she reached mission control, *Soyuz 1* was tumbling through space. Komarov had become ill from the violent motions several times. He forced calm into his voice when flight control told him he could talk privately with his wife.

They brought Valentina to a separate console and moved aside to assure her privacy. In those precious moments Vladimir Komarov bid his wife good-bye.

Soyuz 1's retro-rockets slowed his speed, but reentry began with Komarov having little control. He fought the spacecraft with his experience, skill, and courage. He judged his position by gyroscopes in the cabin. Incredibly, he aligned the ship correctly and held it firm until building atmospheric forces helped stabilize it.

First reports of the ship's landing indicated it had touched down about forty miles east of Orsk. It seemed Komarov had accomplished the impossible, fighting his failed spacecraft all the way through reentry.

But the flight controllers had not witnessed what the farmers in the Orsk area saw as Soyuz fell to Earth. Though Komarov survived reentry, he was fighting his failed spacecraft spinning wildly all the way down.

The main parachute had not fully opened. His reserve chute fell away from his spacecraft, immediately twisting into a large, orange-and-white rag, trailing uselessly behind Soyuz.

At a speed of 400 miles per hour, cosmonaut Komarov's spacecraft smashed into the earth. Its landing system included braking rockets fired normally just above the ground to cushion touchdown. But Soyuz slammed down hard and the rockets exploded, engulfing the spacecraft in flames. Farmers ran to the ship to throw dirt on the burning wreckage. An hour later they were able to dig through the smoldering ruins to find the remains of spaceflight's latest hero, Vladimir Komarov.

The Russian space program, like its American counterpart, had experienced a stunning reversal and entered a period of reexamination. Russia as would the United States did not fly another spacecraft for 18 months.

Neil Armstrong flying the LLTV. (NASA)

THIRTEEN

———

HOW TO LAND ON THE MOON

Following its launchpad fire Apollo would be grounded for 21 months, time needed to redesign and build new, safe flight hardware, time Neil knew the astronauts should be training to fly what was being rebuilt.

Planners came up with a free-flying mechanical contrivance that duplicated lunar gravity and flew much like experts anticipated the lunar module would. It was named the Lunar Landing Training Vehicle (LLTV), and Neil simply reasoned landing a spacecraft in the vacuum surrounding the moon required something altogether different from the skills used to fly winged vehicles in an atmosphere. Flying helicopters could not mimic the trajectories and sink rates of a lunar module and he felt the helicopter would be a waste of time.

The LLTV was a beastly looking contraption, an overgrown spiderlike vertical takeoff vehicle with four legs. It was immediately nicknamed the "Flying Bedstead."

In its stomach was a 4,200-pound turbofan engine mounted on a gimbal, a device that allowed the engine's thrust to swivel freely in all directions, keeping the LLTV balanced, which permitted its jet engine to be throttled down to support five-sixths of its weight to simulate the reduced gravity on the moon.

Why five-sixths? Because Apollo pilots would be landing on the moon in one-sixth the gravity of Earth, they had to know how the lunar module would fly in this low gravity, and to simulate these conditions, the LLTV had two hydrogen peroxide lift rockets with thrust that could be varied from 100 to 500 pounds. Sixteen smaller thrusters, mounted in pairs, gave the astronaut control in pitch, yaw, and roll, and for safety, there were six 500-pound-thrust backup rockets that could take over if the big turbofan engine failed, including one of the first zero-zero ejection seats.

So what is a zero-zero ejection seat? The two zeros refer to zero altitude at zero speed. The ejection unit can boost a pilot to safety from a vehicle sitting on the ground or from thousands of feet in the air. All of it was needed to train an astronaut to fly Apollo's lunar module, and Neil was convinced that if you could build a trainer that would accurately duplicate all the landing conditions on the moon, you could train someone to touchdown safely on Earth's nearest neighbor.

"Those smart Bell aero systems engineers did just that," Neil said, adding, "The only thing they couldn't do was rid the Earth from training in the wind because there is no wind on the moon."

There were other questions but Neil thought the LLTV was just one dandy idea so after the *Apollo 1* fire, NASA's prime Apollo contractors North American and Grumman got busy rebuilding and fireproofing the Apollo and lunar module while the ships' likely moon-landing commanders Neil Armstrong, Pete Conrad, Jim Lovell, Alan Shepard, David Scott, John Young, and Gene Cernan busied themselves trying to master flying the Bedstead.

The skill necessary was the same eye-and-muscle coordination one needed to balance a dinner plate on a broomstick with one finger.

Then, on March 27, 1967, two months to the day after the *Apollo 1* crew died, Neil made his first and second test flights in the LLTV. But technical problems in the maiden Lunar Landing Training Vehicles grounded the first machines for the remainder of the year.

Early in 1968 improved versions were brought to nearby Ellington Air Force Base from Edwards High Speed Flight Station. Neil was first in line, and within a month his logbook recorded a total of ten training flights. Naturally he was asked if flying the Bedstead was becoming routine. "I wouldn't call it routine," Neil answered. "Nothing about flying the LLTV is routine. It's not a forgiving machine."

He continued his training flights and then shortly after lunch on May 6, 1968, Neil climbed aboard the lunar trainer and took his seat in the pilot's box. This would be his twenty-first LLTV practice landing and he got comfortable. Technicians began the training procedure by connecting Neil's necessary attachments.

It was one of those springtime Texas afternoons with lots of sun and puffy clouds, but this day there were also winds. Gusts were up to 30 knots, and Neil made a mental note. The winds could be tricky. Some of the Apollo astronauts disliked flying the demanding LLTV, but Neil felt differently. He knew if a pilot was going to land on the airless moon, he had to know how to do it before he arrived. He had little appetite for "learning on the job."

Was he afraid? Not really. He simply believed in courage over timidity. He had an appetite for adventure over the love of doing nothing, but he certainly had respect for what could go wrong and he drove himself to the limits of being prepared. When he flew the unforgiving Bedstead or any other challenging machine, Neil consciously or unconsciously came back to the dream—the reoccurring dream from his adolescence where he was suspended in air. No aircraft. No wings. Only himself floating thousands of feet—even miles above in the sky—and as long as he held his breath, relaxed and kept his wits about him, he would not fall.

Strangely, he believed the dream to be real. It gave him confidence. Not a boastful confidence. Not arrogance. Again, a belief in his preparedness—in his acute awareness of how sharp and quick his flying skills had to be to pilot any nerve-racking unforgiving quirks of unproven craft. He simply prepared himself the best he could and lived the life of the test pilot.

Aboard the LLTV, Neil's preflight checkout was normal and thorough. With one last pull on his seat harness he and the control van's four engineers and ground crew were ready. The operations engineer gave the go, and the sound of the deep whining of the trainer's turbofan jet engine spinning faster and faster in its start-up told nearby personnel it was time to move away to safety.

Neil increased power and he felt the LLTV lift. Not unlike a helicopter leaving the ground, but with one great difference: Instant loss of power could quickly allow the lunar trainer to drop like a rock before its six 500-pound backup rockets would take over; on a helicopter spinning blades would make for a softer landing.

The hand positions on the LLTV controls in fact were mostly opposite of those on a helicopter—all strange and different but controls in which Neil had become comfortable.

He settled into his climb, gaining altitude much slower than usual because of the vehicle's fight with wind gusts. Suddenly coming into view was the Manned Spacecraft Center's landmark Building One, dominating the center's campuslike skyline. Quickly he returned his focus to his ascent to 450 feet where he would begin his run.

Reaching his assigned altitude was anything but normal. Normal was a little more than a two-minute climb, but the wind gusts and strong head-winds retarded Neil's ascent greatly. He needed twice the time to get upstairs. Four minutes and 26 seconds in fact to reach 450 feet, where he transferred the LLTV's rockets for their lunar-simulation landing.

He used the automatic control system to set the turbofan engine thrust at five-sixths the vehicle's weight and all appeared ready to the control van's four test engineers.

"You have a good weighing at 450 feet," the operations engineer told Neil.

"Roger."

"Come on down."

"Roger."

Neil moved into his simulated lunar-landing run as he had on twenty previous test flights. He was using both attitude-control systems to offset gust reactions. All pictorial and telemetered data told the ground Neil was making a normal yet rapid and somewhat steeper descent through the so-called "Dead Man's Curve"—the terminal phase—and all involved were convinced it was just another day at the office for Neil Armstrong.

But suddenly it wasn't.

The left "low rocket fuel" light flickered then came on steadily, but because of its location in the extreme upper right of the panel, it went unnoticed by Neil.

For only a short instant, the LLTV settled in a hovering position about 30 feet above ground when Neil immediately sensed something was wrong.

"Come out of the lunar sim," the operations engineer ordered and Neil's instincts were cat quick. Before the operations engineer could repeat, "Come out of lunar sim," Neil depressed the lunar simulation release switch advancing the jet throttle to 100 percent. He was instantly climbing about 45 feet per second—reaching for altitude and safety—when the operations engineer saw the helium low light come on.

"Come on back down. We've got helium low," he ordered.

Neil halted his climb at 4.22 seconds with his vehicle nearing 200 feet. For the next 7.34 seconds the LLTV moved forward and level as Neil reoriented vehicle and self.

He knew the loss of helium pressure would shut down his control rockets and just as quickly as he'd reminded himself of this, it happened. He had no control. The LLTV started a pitch-up attitude and began sagging, rolling to the right. Neil tried to regain control but he knew he was sinking. He knew he needed to save the expensive high-tech machine. Deke had the lunar trainer on a low budget, but he also knew he couldn't roll through 30 degrees. If he did and he fired his ejection seat, the seat's rocket would drive him headfirst into the ground.

"Instantly I was out of choices," he later told me. "I lost the pressure and gas to the attitude-control rockets, and when you lose attitude control it diverges, and there was very little time to analyze alternatives at that point. It was just because I was so close to the ground," he explained, "below 100 feet in altitude— time for instant decision, time to depart—and I told the van, 'I have to leave it.'"

"Leave the vehicle?" the operations engineer questioned.

Neil's hand was already on the tiger-striped ring beneath his seat. He jerked it forward and felt his world explode.

He and his zero-to-zero ejection seat were blown instantly from the LLTV and in a blink of the eye a small explosive blasted his parachute canopy open.

There was no waiting for a drogue. No waiting for the main chute to blossom. Parachute deployment was instant. Rocket power and pressurization fully opened his chute in a second flat.

Following Neil's ejection, the failing LLTV had taken only 2.84 seconds to hit the ground, crumbling and exploding into a violent fireball.

Neil was watching from a safe distance, grateful to be riding his parachute away from the burning fuel and melting metals. Some had feared an ejecting pilot would simply fall into the middle of the burning crash.

Neil was thankful he did not. He judged his chute was taking him about the length of a football field from the burning wreck into a patch of waste-high weeds. He refocused his attention on making a good parachute landing by pulling his knees together, his legs upward, and holding firmly to his parachute straps as his feet parted weeds.

He rolled to a stop on the ground 10.34 seconds after ejecting, pulling in his parachute and harness before taking stock of his body and limbs. He reminded himself that it was his first ejection in seventeen years—since he ejected from his crippled Panther over Korea. With the exception of biting his tongue, he would realize later that night the only other damage he suffered was a bad case of chiggers he got in the weeds.

Other than that Neil brushed himself off and was grateful he hadn't a scratch when help arrived. He was taken back to the staging area for a quick debriefing with the ground crew.

Neil Armstrong ejects from his failed Lunar Landing Training Vehicle and rides his parachute to safety above the burning crash below. (NASA)

The time was 1:45 P.M. and the unflappable test pilot continued dusting himself off all the way to his car. He nursed his lacerated tongue and drove his 'Vette back to his office. He was satisfied he had done what he'd been trained to do and there was no reason to make more out of it.

Neil went back to work, and in the coming days investigators found the cause of the LLTV accident. It was a poorly designed thruster system that allowed propellant to leak out, and the loss of helium pressure in the tanks caused attitude thrusters to shut down. Neil had no control, and did nothing that contributed to the accident, and while applauding Armstrong for his flight skills and decision-making, Manned Spacecraft Center Director Bob Gilruth and Director of Flight Operations Chris Kraft felt it was only a matter of time before an astronaut would be killed in the lunar trainer.

The executives were ready to eliminate LLTV training completely, but Neil, Pete Conrad, Alan Shepard, Dave Scott, and Gene Cernan—five of the six Apollo commanders who would land on the moon—were adamant they needed the LLTV.

"Forget about punching buttons in a safe ground trainer," Neil told the bosses. "Would you have us train for real only once—when we were 200 feet above the moon's surface, or would you rather for us to learn above Earth where we have help?"

The man who would be last on the moon, Gene Cernan, backed the astronaut who would be first, and Cernan didn't sugarcoat his words: "Training with the LLTV means you ain't allowed any mistakes. It puts your ass in the real world. We have to have it."

Neil wasn't easily excited but he believed strongly in the fact that you had to expect some things to go wrong, and you needed to prepare yourself to handle the unexpected. "You just hope those unexpected things are something you have prepared yourself to cope with," he told me.

Some writers wrongly accused NASA of exaggerating Armstrong's ejection altitude of about 200 feet. The writers stated it was more like 50 feet with Neil having only two-fifths of a second before impact. Neil told me he judged he was slightly less than 100 feet, the point where the rules called for the pilot to eject if he was having problems. The video confirms Neil's judgment. Apparently those writers were confusing the facts of the LLTV's first ejection by its test pilot Joe Algranti.

From Neil's mouth to this writer's ears, from the videos, the investigations, and the interviews with others who were there, it was revealed Neil was higher than 50 feet when he ejected and his lunar trainer took 2.84 seconds to hit the ground. Not a scant two-fifths of a second, and most important, there was never an accident in LLTV training caused by an astronaut.

The training continued with Neil making almost sixty landings in the LLTV, and after he had made history landing first on the lunar surface, he said, "The LLTV gave me a good deal of confidence—a comfortable familiarity when we landed on the moon."

Next for Neil was *Apollo 8.*

Deke Slayton had selected him as backup commander to Frank Borman, and Edwin "Buzz" Aldrin as the backup command module pilot to Jim Lovell. But Mike Collins, who would fly to the moon with Neil and Buzz on *Apollo 11*, had to stand down for a neck operation. Deke moved astronaut Fred Haise in as the third member of the *Apollo 8* backup crew.

The mission plan called for *Apollo 8* to follow the first Apollo crewed mission into Earth orbit, and that flight, dubbed *Apollo 7,* was on the calendar for October—less than seven months away.

Fourteen months had passed since the *Apollo 1* fire and *Apollo 7*'s scheduled launch seemed promising. Neil took stock of the fact that half a billion dollars had been spent on the exhaustive redesign and rebuilding of Apollo, including a new hatch that an astronaut could open in three seconds flat. The new spacecraft included extensive use of fire-resistant materials, a redesigned electrical system, better protection for plumbing lines, and use of a combination nitrogen-oxygen atmosphere system when the spaceship was on the ground.

Neil was aware that what happened to Apollo could also happen to the lunar module, a fact not lost on its builder Grumman Aircraft. The Grumman team made extensive changes to that vehicle, too, and although Neil didn't know it at the time, Grumman was building the only space vehicle that would carry humans without ever suffering a failure. It would in fact be the lifeboat that would later save the three *Apollo 13* astronauts.

The Grumman team quickly earned the respect of the industry and Neil, remembering his friend Gus Grissom, who hung a lemon on the *Apollo 1* simulator, spoke often about the efforts before assemblages.

Neil would say, "I'm convinced we would have ended up losing more lives in a number of ways before we got to the moon, and we may never have gotten there if it hadn't been for *Apollo 1*. We uncovered a whole barrel of snakes. We would have fixed them one by one. The fire forced us to shut the program down and redo it right, and we got there on the backs of Gus, Ed, and Roger."

On October 11, 1968, the Saturn 1B rocket thundered from its launchpad and boosted *Apollo 7* and its crew of three into orbit.

Halfway through powered flight Commander Wally Schirra keyed his microphone and told Mission Control, "She's riding like a dream."

From launchpad 34, where Gus Grissom, Ed White, and Roger Chaffee died, their backups Wally Schirra, Walter Cunningham, and Donn Eisele leave for Apollo's first flight. (NASA)

Apollo 7's Schirra, Walter Cunningham, and Donn Eisele spent eleven days in space. They tested the new Apollo systems, conducted experiments, and beamed the first extensive live television scenes from space to fascinate audiences around the world while proving the new Apollo would work in space longer than needed for a flight to the moon and back.

The *Apollo 7* astronauts gave us this extraordinary picture of sunrise over Cape Canaveral and the peninsula of Florida. Some say it was the gods' memorial for the fallen *Apollo 1* crew. (NASA)

America was back on the road to the lunar surface, but Russia wasn't standing still. America had its super booster the Saturn V. Russia had its super booster N-1. Saturn V was on schedule; N-1 was beset with failure.

America's intelligence assets were keeping a close watch on the Russians

Russia's Zond spacecraft on its flight around the moon. (Artist image, Russian Federal Space Agency)

and Neil had friends in those assets. He was one of the first to be told when the Russians prepared a squadron of their smaller R-7 rockets. They were to launch a fleet of tankers into Earth orbit. Cosmonauts would follow and gather the tankers and herd them into a single train of one of the all-new Soyuz spacecraft and five fueled cars. This train would chug its way to lunar orbit, beating Americans to the moon.

Then Vladimir Komarov died in *Soyuz 1*, delaying Soviet plans.

The Russians regrouped. They began modifying the Soyuz spacecraft so it could carry one or two cosmonauts on a single pass around the moon. Not a landing. But in the eyes of most nations they would have indeed beaten the Americans to the moon and then gone on to other projects such as building a space station, saying, "We could've landed if we'd wanted."

The Russians called their modified lunar ship Zond, and they loaded it with

tortoises, flies, and worms, and in November 1968, they sent the living creatures on a flight around the moon and brought them all back alive.

Neil and his friends in intelligence believed this flight was Russia's dress rehearsal for history's first trip by a cosmonaut around the moon and back. NASA
did, too.

Sitting atop its mammoth Saturn V rocket, *Apollo 8* is rolled out of its giant assembly hangar for its three-and-a-half-mile, snail-pace journey to its launchpad. (NASA)

FOURTEEN

———

HELLO MOON

A Thor-Agena rose from California's Vandenberg Air Force Base carrying another photoreconnaissance spy on its mission of keeping watch on what was happening at Russia's launch site. Come the following morning, analysts were looking at another Soviet heavy-lift proton rocket sitting on its launchpad. It was waiting to be outfitted with a Zond spacecraft.

Time was critical.

As they had done with the Zond loaded with tortoises, flies, and worms, the Russians were now in a position to dispatch a single cosmonaut on a circumlunar flight in December or January.

NASA's top executives were in a quandary. They saw failure before them.

NASA administrator Jim Webb told President Lyndon Johnson it was time for America to gamble, to consider putting astronauts on the Saturn V rocket's first manned flight, and send them all the way to the moon aboard *Apollo 8*.

Some experts scoffed at the plan. Most argued for it, and Webb told the outgoing president that it was the consensus of NASA engineers that they had corrected the Saturn V's minor problems and there was no need for an additional unmanned test. It was time to fly.

America wasted time with a test of the Redstone that proved to be useless, which permitted Yuri Gagarin to beat Alan Shepard into space. Deke Slayton and NASA's astronauts weren't about to see a repeat of that little dog and pony show. The lunar module would not be ready for its first flight test for four or five months, but they had a perfectly good Apollo, and Deke turned to *Apollo 8* Commander Frank Borman, later saying, "The sonofabitch almost turned handsprings when I told him there was a possibility *Apollo 8* would go all the way to the moon."

Deke laughed. "Borman's answer was an overwhelming *yes*," and then Deke told *Apollo 8*'s backup commander, Neil Armstrong.

Neil assured the boss that with the success of *Apollo 7* the idea of sending *Apollo 8* to the moon was a masterful stroke of genius. He had been talking to his friends in intelligence and Neil told Deke, "We should not only go, we should put *Apollo 8* in orbit around the moon, too. This would kill their plans to fly circumlunar."

Deke nodded. He liked Neil's advice.

Every second of time was essential. On November 11, 1968, the new NASA administrator, Thomas Paine, approved the plan. He phoned his decision to the White House, and President Johnson gave his blessing. It was the single greatest gamble in spaceflight then, and since.

Apollo 8 was readied for launch December 21, 1968, and Neil awoke with the prime crew. Frank Borman, Jim Lovell, and Bill Anders began suiting up as Neil hustled over to the launchpad and climbed aboard the moonbound spacecraft. It was the job of the backup commander to monitor the prelaunch sequence from inside the cockpit. He was there to check and set all switches.

When the suited prime crew arrived in the white room, Neil shook their hands and retreated to the Launch Control Center where he joined his backup crew, Buzz Aldrin and Fred Haise. The three staked out great viewing spots along the big window facing the pad and turned their ears to the speakers and the voice of launch commentator Jack King.

"Ten, nine, eight, seven, six," and the Saturn V was alive.

But it didn't go anywhere. Neil knew the giant rocket consisted of millions of parts and systems. The Launch Control Center's computers worked at the speed of light checking and rechecking every single part before the most powerful machine ever would be permitted to move.

But its sound didn't stay put.

Its thunderous roar came to life with Saturn V's ignition and tore its way across the space center, hammering everything in its way. The wide launch control building was no exception.

The tsunami of thunder slammed into the nerve center's windows, buckling the big one in front of Neil and crew, who for a moment thought they'd bought it. They thought all glass had shattered and they stepped back as newly built ceiling parts fell along with other construction leftovers.

Those in Launch Control thought the sky was falling. (NASA)

But the windows held and Neil could easily see the Saturn V's powerful engines were burning even fiercer, demanding they be unleashed, and Jack King reported, "All engines running . . . Three, two, one, zero!"

And then it happened. Explosive bolts fired and the Saturn V's giant holddown arms released their grip.

"We have liftoff, liftoff of *Apollo 8*—destination, moon."

Neither Neil nor any among the huge assemblage surrounding the launchpad could take their eyes off the enormity of it all: Saturn V moved. That largest of machines ever created reached for sky, rode on flame and roared, pounded ears, overwhelmed all those watching, slammed crackling thunder into their bodies, fluttered their clothes, rolled their flesh in small yet perfect patterns,

Apollo 8 heads for the moon. (NASA)

The 600,000 who had gathered on and around the space-port braced for the Saturn V's fury. (NASA)

and rattled the coins in their pockets. Neil Armstrong suddenly knew he couldn't wait for the day he'd be riding that beautiful son of a bitch.

Birds fled from their roosts. Wildlife ran from the stunning and numbing sound. It pounded and leapt and trampled until it was no longer thunder, no longer roar. It turned into a series of staccato explosions and now it hurt. It brought a terrible crackling pain to the ears, assaulted the body, yet it was exhilarating and worth the beating as the great assemblage stared into the blinding mass of fire.

Higher and higher *Apollo 8* climbed, leaving its ear-shattering sound behind as it reached for orbit on a spear of flame more than 800 feet in length. The mass of spectators could only stare deeply into its flaming thrust, watch it

Apollo 8's Saturn V sends birds and wildlife fleeing. (NASA)

turn into a rich orange, watch as red appeared along its burning edges, and each sought a final last sighting as the pounding chariot drawn by thrust and driven by fire disappeared over the Atlantic and Neil Armstrong knew that if one could love a machine he loved this one.

On board Frank Borman, Jim Lovell, and Bill Anders had just pushed through the region of maximum dynamic pressure, their spaceship leaving all sound behind. It was eerily quiet now. Had the astronauts not heard the humming of electronics, they might have thought they were in a simulator on the ground.

But they sure as hell weren't.

Apollo 8's astronauts leaving for the moon. (NASA)

Something was slamming their growing weight back into their seats, haul-ing them faster and faster into space, and it was all working. By the time they neared the speed needed to orbit Earth they had burned and discarded the rocket's first and second stages. Now all they needed was a push from its S4B third stage to begin circling Earth, which they got, and Mission Control be-gan checking and rechecking all of *Apollo 8*'s systems again. They had to be as sure as possible it had survived the rigors of launch and its ride into orbit and all would be well for the farthest and the longest journey ever taken by humans.

The checks and tests lasted for nearly two Earth orbits. *Apollo 8* was set. The crew then fired the S4B stage again.

The final burst of energy from their Saturn V's final stage increased the astronauts speed more than 7,000 miles per hour. The rocket burn was what space navigators called translunar insertion, and *Apollo 8*'s new speed of more than 24,000 miles per hour broke the grip of Earth's gravity. The spaceship with a crew of three was free to cross the void to the moon. Shortly thereafter Jim Lovell sent a message back to Mission Control. "Tell Conrad he lost his record."

Pete Conrad and Dick Gordon had reached a height of 739 nautical miles in their agile *Gemini 11* spacecraft. But *Apollo 8* was still climbing and when it was done the new record would be some 240,000 miles.

Inside Launch Control Neil had monitored every happening on board the moon-bound ship. But now it was time for him and the backup crew to get moving—back to Mission Control to witness *Apollo 8* become the first manned spacecraft to enter an orbit around the moon. He and Buzz and Fred with their wives boarded a NASA Gulfstream aircraft for the flight back to Houston. When home Neil showered, changed his clothes, and then drove to Mission Control to stand by if needed.

He was there in his capacity as *Apollo 8*'s backup commander. If any of the flight controllers needed to know what the crew was scheduled to be doing at any given moment, he was their man. Otherwise he took a stay-out-of-the-way seat in the big room full of consoles.

Deke Slayton saw Neil and thought this would be a good time to talk about his next assignment. He came over and pulled up a chair.

"Got a minute, Neil?"

"Sure, Boss, anytime."

"Been thinking about your next assignment."

"That's great."

"There's lots of ifs, ands, and buts," Deke said flatly, "but we're thinking about you commanding *Apollo 11*."

"That wouldn't make me mad," Neil grinned.

Deke leaned forward and in an almost whisper explained there was no way of knowing what *Apollo 11*'s mission would be. But, if *Apollo 8*'s current flight to orbit the moon was a success, if the lunar module could pass muster in its Earth-orbital first flight with *Apollo 9*, and if *Apollo 10* could return to lunar orbit and its lunar module could descend to within 8.4 miles of the moon, then *Apollo 11* could be the first to land.

Neil wasn't easily stunned, but he was for a long moment. He just looked at the director of flight crew operations and processed everything he had been told. "Thanks Deke, thanks for your confidence," he said offering him his hand. "If you decide to trust me with *Apollo 11*, you'll get my best effort."

"I know we will, Neil," he assured him, adding, "It's shaking out that way."

Deke walked away and Neil reflected on the stunning possibility his crew just might be the first to set foot on a place other than Earth. But certainly not lost on Neil, despite what he'd just been told, was the fact that the marvelous product of science, technology, and engineering that would take him and his crew there was at this very moment moving between Earth and the moon—the Block II Apollo command module created from the ashes of *Apollo 1*. It was crammed with the knowledge learned from NASA's astounding misjudgment.

The three astronauts riding within the cone-shaped, tiny world of their ves-

For the first time humans were seeing the Earth as an almost full sphere. (NASA)

sel were doing so under the watchful eyes of the largest audience in history. More than a billion were tuned to their radios and televisions receiving reports of sights never before seen.

Live views from a receding Earth came into homes from the spaceship, and its crew played tour guide.

Meanwhile on the steppes of Kazakhstan Zond was left standing on its launch-pad. Disappointment replaced the usual holiday round of vodka and cognac toasts.

Lev Kamanin, top aide to Kremlin space officials and the son of the chief of cosmonaut training, wrote in his diary: "For us this day is darkened with the realization of lost opportunities and with sadness that today the men flying to the moon are named Borman, Lovell, and Anders, and not Bykovsky, Popovich, or Leonov."

No matter how well he said it, Kamanin's sentiment was not the end of Russian efforts to reach the moon. Despite failures with their big rockets, the cosmonauts would continue to try.

The hours in *Apollo 8*'s flight clicked away—day one and now day two—but not the awe and wonder. Armstrong knew the astronauts were fascinated, unable to take their eyes off of what was out there. Suddenly they were aware of a distant sphere easing into view—a stunning view of Earth. Not vast horizons curving gently away, but a more than half-full Earth—a blue marble with dominating blue seas and white clouds, bountiful rain forests and mountains rising above its surface.

From midway between their home and the moon, Earth appeared to them as perfectly round, a stunning sphere, and *Apollo 8* rolled on with Earth sliding silently out of sight leaving the astronauts a universe that was for the first time on their trip totally dark. No more Earthglow, no more moonglow—even the sun had hidden itself—and Borman, Lovell, and Anders had the most clear, the most distinct view possible, of their own Milky Way and hundreds more galaxies with their nebulas and star clusters, some so bright and far away.

Tomorrow would be Christmas Eve. That Borman, Lovell, and Anders were approaching the moon seemed impossible. They had left Earth riding America's largest rocket, the mightiest energy machine ever built to lift straight up and away from the deep gravitational well of their planet, a monster of steel and ice

Apollo 8's astronauts saw this stunning view of Earth with white clouds and dominating blue seas. (NASA)

and fire atop which no man had ever before flown. And they were risking everything to fly into orbit around Earth's natural satellite.

It was a gamble like few others known to history.

Apollo 8 slipped across the equigravisphere, that point in their flight where the moon's gravity would have a greater pull than distant Earth. In Mission Control Neil and Deke Slayton grabbed some fresh coffee and lost themselves in one of the control center's back rooms.

"Mike Collins has recovered well from his neck operation," Deke began, "How would you feel about having Collins and Buzz Aldrin as your crew?"

"No problem," Neil assured him. "I've been working these last months with Buzz on the backup crew for *Apollo 8*, and everything went well."

"No problems?"

"None."

"I could make Jim Lovell available."

Neil was taken a bit aback. He was aware some had difficulty getting along with Buzz, but not him. And there were none better than Jim Lovell and Mike Collins and that begged the question, Which would fly which? Would Mike still be command module pilot and Jim land on the moon or would Lovell, the most experienced command module pilot in the corps, remain in that slot leaving Mike Collins to move over as lunar module pilot?

Neil asked Deke, "Isn't Jim Lovell commander material?"

"Yes, definitely," Deke agreed.

"I wouldn't want to interfere with Jim commanding his own flight."

"Understand—just wanted to give you options," Deke told him. "Let's both think about it overnight."

"Good idea, Boss," Neil nodded, quickly reminding himself that *experience* should be his chief consideration for a crew.

Buzz Aldrin had nailed the EVA thing in *Gemini 12* and Mike Collins was at the top of his game in flying the Apollo command module, but at this moment Jim Lovell was at the helm of *Apollo 8*'s command module.

Later, Neil would tell me, "If we switched things around too much, getting other people's nose out of joint because we stole somebody from somebody else's crew, we could have made enemies.

"And there were personalities—always personalities, and that's one where I came up short," Neil laughed, taking a good-natured jab at himself.

With the moon getting closer *Apollo 8*'s astronauts were waiting for Mission Control's decision on whether or not they would slide into lunar orbit.

The three astronauts did not know if they would be spending Christmas circling the moon or making a single trip around its far side. Either way their flight was already a smashing success. It had opened the door to human exploration of our solar system, but, and it was a big but, everyone in NASA as well as the world's billions in the radio and television audience wanted *Apollo 8* to go all the way.

Neil knew in the astronauts' service module was the SPS, *Apollo 8*'s largest rocket. It would be needed to reduce their speed to place them in lunar orbit, and then would ignite again to bring them safely home. The question in Mission Control was, "Is lunar orbit the safe thing to do?"

Neil also knew what the decision-makers must consider was critical not only to the mission but also to the lives of the three men. To slip *Apollo 8* into lunar orbit, the big SPS had to fire at full thrust for precisely 247 seconds. Following shutdown, the crew would use its attitude-control thrusters to point the nose of its ship in the direction of flight. If the engine burn faltered or failed early, the astronauts would soar past the moon on a path that would not return them to Earth. If the rocket burned too long, *Apollo 8* would crash somewhere on the lunar landscape.

If SPS failed to ignite altogether, or if Mission Control decided not to go for lunar orbit, *Apollo 8* would be perfectly safe. It would swing around the far side of the moon, curving in its sharp orbit as if it were a celestial boomerang and, without using an ounce of rocket fuel, the astronauts would be on their way home. This was the "Free Return Trajectory" inserted into *Apollo 8*'s computers before it left Earth.

Neil leaned back in his chair knowing this was Mission Control's "moment of truth." He also knew all the control center's monitors were "green."

Astronauts Frank Borman, Jim Lovell, and Bill Anders were ready for any emergency. They were about to disappear behind a 2,000-mile-wide celestial body— fly across the side of the moon facing away from Earth where signals between *Apollo 8* and Mission Control would be blocked for more than twenty minutes.

CapCom Jerry Carr received the nod. It was his job to give the astronauts their answer and everyone crossed his or her fingers to hear. "Ten seconds to go," he told *Apollo 8*. "You are *Go* all the way."

Jim Lovell's voice was incredibly calm. "We'll see you on the other side, Houston."

With those words *Apollo 8* vanished behind Earth's closest neighbor.

Neil listened. But there was nothing to hear. The mission had simply gone quiet. No communications. No telemetry signals. No way of knowing if the three astronauts continued to exist.

What Mission Control couldn't know until *Apollo 8* emerged from behind

With the moon closer to them than Earth, the astronauts had a crater-potted lunar surface staring them in the face. (NASA)

the moon was at the precise moment dictated by its flight plan, Jim Lovell had fired *Apollo 8*'s biggest rocket for 247 seconds, a time he would later call the "longest four minutes I've ever spent."

It was a splendid, epochal moment sixty-nine hours and fifteen minutes after launching from Earth, and when the rocket burn was completed, *Apollo 8* had locked itself into lunar orbit.

Mission Control only knew it should have happened, and the flight controllers continued their cliff-hanging suspense, counting the minutes and seconds before *Apollo 8* would emerge from the other side of the moon.

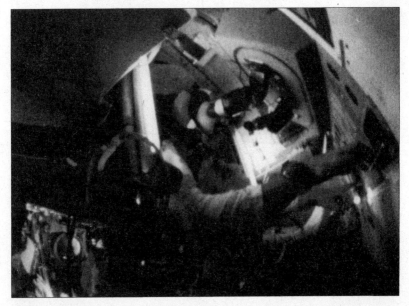

Jim Lovell fires *Apollo 8*'s largest rocket engine and places humans in orbit around the moon. (NASA)

CapCom Jerry Carr could only keep up his persistent call, *"Apollo 8 . . . Apollo 8 . . . Apollo 8 . . ."*

It seemed like an eternity but then the intense clock-watching was over. Headsets and speakers crackled, and Neil heard the voice of Jim Lovell calm as always, "Go ahead, Houston."

Those three words—coming just at the instant they should have—sent Mission Control into a bedlam of cheering, whistling, shouting, and back-slapping.

Apollo 8's telemetry flashed numbers on the big viewing board. It was in an orbit 60 by 168.5 miles above the moon. Later, on the third trip around the lunar surface, the ship's big rocket fired again and dropped the astronauts into the desired, nearly circular orbit of 60.7 by 59.7 miles.

But the thrilled global audience didn't want numbers. It wanted to know what the moon looked like.

"Essentially grey, no color," reported lunar tour guide Jim Lovell. "It's like plaster of Paris or a sort of greyish beach sand."

In their first two telecasts, the astronauts transmitted video from lunar orbit of the wild and wondrous landscape pitted with massive craters. "It looks like a vast, lonely, forbidding place, an expanse of nothing," said Borman as Lovell saw the distant Earth as a "grand oasis."

The third member of the crew Bill Anders added, "You can see the moon has been bombarded through the eons with numerous meteorites. Every square inch is pockmarked."

Lovell added, "The vast loneliness is awe-inspiring, and it makes you realize just what you have back there on Earth."

That Christmas Eve in 1968 was extraordinary not just for Neil and Deke and the others in Mission Control, but for the billions that had been brought together before their television sets. They were seeing wondrous never-before-seen video of

Once in orbit around the moon, one of the first things seen by *Apollo 8*'s crew was Mother Earth rising above the lunar landscape. (NASA)

the moon moving quietly below *Apollo 8*'s lunar orbit when Bill Anders spoke: "For all the people on Earth," he began soberly, "the crew of *Apollo 8* has a message we would like to send you." He paused briefly and began reading from the verses of the book of Genesis: "In the beginning, God created the heaven and the Earth . . ." As Bill concluded the fourth verse, Jim Lovell read the next four with Frank Borman concluding with, "And God called the dry land Earth, and the gathering together of the waters He called seas. And God saw that it was good."

Earth, the place God took his time to make. (NASA)

The moon with its view of the distant, soft blue marble of life had become host to poets, and Borman signed off with, "And from the crew of *Apollo 8*, we close with good night, good luck, a Merry Christmas, and God bless all of you—all of you on the good Earth."

Neil clutched his emotions thinking, *I hope my Mom saw this.* She would have enjoyed it he told himself rising from his chair.

His thoughts returned to the questions of the moment. If what he and Deke had been talking about came to be, he'd be there soon, possibly on the moon itself, and Neil went off to find Deke. They had agreed on a final meet.

The two found an out-of-the-way niche, again in the back of Mission Control. When each had their say Neil was content that Mike Collins and Buzz Aldrin would serve with him on *Apollo 11.* Jim Lovell, Bill Anders, and Fred Haise would be their backups.

"I'll be announcing the crews next month," Deke told Neil, and when I saw him later, Neil told me off the record he would likely command *Apollo 11.*

"Doesn't surprise me a bit," I nodded, smiling. "You are obviously the most qualified pilot. But possibly more important, you have the character and training to handle it." "Maybe," he said. "It was the luck of the draw."

I knew this man standing before me never lobbied to make the first landing. He never sought any more consideration than his fellow astronauts, and I could only stare at him. I was amazed this humble man truly believed he was only selected because he was next in line.

I understood why the other astronauts believed that, but there was no question in my mind that the engine driving this selection was the NASA bosses' awareness of Neil's devotion to flying the Lunar Landing Training Vehicle until he had this landing on the moon thing nailed along with his incessant need to know what to expect. He simply had to be prepared for the unexpected.

"This is off the record," I told Neil. "As you know Deke and our man Harold Williams are fishing buddies, and he told Harold straight out he wanted you first, and if you had to abort he knew Conrad could handle the job on *Apollo 12,* or if not Conrad, Jim Lovell could get it done on *13.*"

Neil gave me a little smile that said maybe yes or maybe no, and I stood there admiring the raw fairness of this man. I told him sincerely, "Neil you are too good. You are the most considerate person I've ever met, and you may not think you're special but everybody else does. And dammit," I raised my voice, "if you ever need me to run interference for you just let me know."

Neil nodded and gave me the greatest compliment of my life. "I'd like you as my blocking back anytime," he smiled.

"Just ask," I said knowing the best part was that he was sincere.

Early in the morning of Christmas Day, *Apollo 8* moved through its tenth and final orbit around the moon and was again out of contact with Mission Control. The critical rocket firing would either start them on their journey home or leave them stranded in lunar orbit. At the appointed moment, Borman, Lovell, and Anders felt their big SPS rocket come to life, creating a long stream of flame and a wide plume of fire behind the engine. On the 304th second the engine shut down, right on the mark.

Time dragged maddeningly for a waiting world.

Finally, Jim Lovell's voice came through with pure joy: "Please be informed there is a Santa Claus," he said. "The burn was good."

It was better than that. After twenty great hours in lunar orbit, *Apollo 8* was driving right down its own pioneered Earth to moon interstate, down the mathematical highway it had to fly to reach a point 400,000 feet above Earth— the exact angle and altitude to reenter with a speed greater than any human had ever flown.

Two-and-a-half-days after leaving the moon, Earth's gravity reached out and dragged Borman, Lovell, and Anders back into its atmosphere with *Apollo 8* becoming a man-made meteor. Temperatures soared to those on the surfaces of stars, and plunging downward the astronauts knew their lives depended on how well their ship had been built.

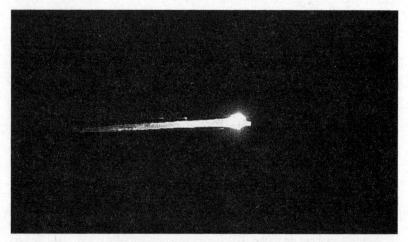

A telescopic camera caught this picture of *Apollo 8*'s reentry into Earth's atmosphere. Only fire could be seen, blinding white flames within a red sheath leaving a burning trail 125 miles long. (NASA)

Apollo 8 traded its speed for heat. The hotter the fire flowing from the heat shield the slower the spacecraft, and suddenly they were safely two miles above the Pacific, in sight of Christmas Island, and three large parachutes streamed away, blossoming wide and full.

The world cheered.

The astronauts returned to a thundering ovation.

The road to the moon was open.

Neil Armstrong, Mike Collins, and Buzz Aldrin: the misfits at the roll-out of their *Apollo 11* Saturn V rocket. (NASA)

FIFTEEN

———

THE MISFITS

The talk around NASA was the first "moon landers" had been selected. All eyes focused on Deke Slayton, and on January 4, 1969, he called Mike Collins and Buzz Aldrin into his office. Neil was already there.

"I'll get right to the point," the director of flight crew operations began. "Neil will be commanding *Apollo 11*, and we'd like you, Mike, to be the command module pilot." He shifted his eyes to Aldrin. "Buzz we'd like you to handle the lunar module and shepherd the experiments.

"It's conceivable you guys could make the first landing on the moon," he told them soberly. "We want you to train that way."

Collins and Aldrin grinned like Tennessee mules eating briars as stoic Neil Armstrong, long a devotee of Zeno of Elea, a pupil of Parmenides who was unmoved by joy or grief, stood without expression. Neither Collins nor Aldrin had a clue Deke and Neil had been meeting during *Apollo 8*.

"Mike, you, Jim Lovell, and John Young are at the front of the line when it comes to handling the command module," Deke continued. "You'll be taking care of your crew's only ride home," he said grinning, "and if I were Neil and Buzz I'd keep you happy."

They all laughed and Mike said, "I'll keep the home fires burning, Boss."

Deke nodded, turning again to Aldrin. "Buzz, your spacewalk on *Gemini 12* was a classic, rid us of lots of problems. That was a great job, and we're convinced you can do the same as *Apollo 11*'s lunar module pilot."

"You'll get the best I have, Boss," Buzz nodded.

"Good, we want you to know every bolt and washer in that LM [pronounced Lem], and if it gets a gut ache, I want you to be ready with the Alka-Seltzer."

"I will, Deke," Buzz assured him once again.

"As commander Neil will make all the decisions, asking of course for your input. He'll have a full working knowledge of both ships and Neil has come a long way in mastering the landing of a lunar module primarily with the LLTV, and Buzz," Deke cleared his throat, "we want you to use that MIT sheepskin of yours to please the science boys with the experiments. You'll be setting them up, and the science guys would like for them to work for decades—we want them chirping back their data to every interested scientist on Earth."

"Yes, sir."

"We won't let you down, Boss," said Mike, adding his own assurance.

"It's going to be a great flight whatever the mission turns out to be," the newly appointed *Apollo 11* commander added. "We'll be ready, Deke."

The director of flight crew operations offered his hand for each of them to shake, believing he had a solid crew to make the first landing on the moon.

The NASA brass was in full agreement with Deke's decision. The boys at the top—considering the era they were all boys—were sure that if Neil were the first on the moon he would not cheapen that honor by enriching himself. The world would not see a chain of "Neil Armstrong's Moon-Burger Drive-ins," or toothpaste endorsements. This American son, who began earning his own way at age ten, would honor the historic gift with deserving dignity. This alone, disregarding the fact that Neil Armstrong was arguably the best pilot in the astronaut corps, was reason enough to give him the job. If problems cropped up and *Apollo 11* had to be waved off, *Apollo 12* had a commander who wasn't too shabby himself. His name was Pete Conrad and no one would argue Pete couldn't fill Neil's shoes—certainly not Neil. Either, along with the likes of John Glenn, James Lovell, Alan Shepard, Dave Scott, John Young, and Gene Cernan would serve America well as its first ambassador to a place other than Earth. No one was more aware of this fact than Neil, who would spend his life believing any one of his astronaut brethren could have performed as well as he did.

NASA announced the *Apollo 11* crew five days later following White House ceremonies for the *Apollo 8* astronauts. Outgoing president Lyndon Johnson awarded Frank Borman, Jim Lovell, and Bill Anders medals and the *Apollo 8* crew received standing ovations at a joint meeting of Congress.

Neil was pleased with Mike's and Buzz's selection even though many in the space family regarded the *Apollo 11* crew as the "Misfits."

Socially that description may be true. Neil and Mike and Buzz weren't pals. They saw little of each other while not at work; they even drove their own separate cars to the job. But Neil knew when it came to Apollo's command module there was no one out there who could best Mike Collins. The same was true when it came to Buzz Aldrin. Who was going to compete with Buzz in the smarts department? He graduated third in his West Point class. He was an F-86 combat pilot in Korea with a couple of kills. Those pilots were simply the best, and Buzz received a doctor of science degree in astronautics from MIT, the Massachusetts Institute of Technology. Who better for the lunar module and setting up experiments on the moon?

Neil's roots were in rural Ohio—well planted with small-town comfort, security, privacy, and most important to him proven values.

This was also true for the original Mercury Seven astronauts.

As John Glenn put it, "Growing up in a small town gives kids something special. They learned how to make their own decisions, and maybe," Glenn added, "maybe it's no accident that people in the space program, a lot of them come from small towns."

"The small towns I grew up in were slow to come out of the Depression," Neil said. "But we weren't deprived. My father's annual salary was about $2,000, and we never had much money around. But to some of my friends, the fact my father had a job meant that the Armstrongs were rich.

"I got my first job when I was ten. I was paid only ten cents an hour. I was happy to get it. I cut grass at the Upper Sandusky's historic Old Mission Cemetery, and I never had the first complaint that I was only ten from its occupiers," he laughed, "and even though I had to cut grass ten hours to make a dollar, I was the only boy around with a dollar."

Mike Collins was the sort of man Neil naturally enjoyed rubbing elbows with—a good-humored man who enjoyed a joke while being thoughtful and articulate and learned. Neil knew Mike was born in Italy and for the first 17 years of his life, he called Rome; Oklahoma; Governors Island, New York; Puerto Rico; San Antonio, Texas; and Alexandria, Virginia, home. This alone settled the question of who was the most cosmopolitan member of the *Apollo 11* crew, the one who often kidded Neil about being from a small town, which Neil countered by telling Mike that those who live out among the cows and chickens think that people who live in crowded streets and the hustle and bustle are the ones with the problems. Neil would add that anyone who lives elbow to elbow with thousands of others was missing the good sense and judgment to come in out of the rain.

As a son of an army attaché in Rome, Mike Collins was born on Halloween October 31, 1930, to Virginia Stewart Collins, a cultured, educated woman, and to Major General James L. Collins, a man who fought with General John J. Pershing in the Philippines, and again in 1916 when the Mexican Revolutionary Pancho Villa raided Columbus, New Mexico. Mike's father and General Pershing chased Villa into Mexico, tracking him for seven months only to be called back with the outbreak of World War I.

But it was Mike's uncle who became the better known. J. Lawton Collins, "Lightning Joe" as he was called, was one of General Dwight David Eisenhower's field commanders in World War II. General "Lightning Joe" would later become chief of staff of the United States Army from August 1949 to August 1953, serving as the Army's senior officer throughout the Korean War while Mike's older brother, James L. Collins Jr., graduated from West Point and became a field artillery battalion commander in World War II. He won a boatload of medals— the Purple Heart, Bronze Star, Silver Star, Distinguished Service Medal, and Legion of Merit. In 1965 Mike's brother would become a brigadier, later a major general, and after Mike Collins completed his tour as an astronaut, he would carry on his family's tradition in the military by receiving stars for his shoulders, too. Before he retired from the Air Force Reserves he received the rank of major general.

As a boy Mike's two older sisters, Virginia and Agnes, mostly ignored their younger sibling who would finally know what an extended homelife felt like with the beginning of World War II. That's when Mike's family moved to Washington, D.C., where they lived for the duration of the war.

The skinny athletic 12-year-old attended the Episcopal preparatory school, St. Albans, where he captained the school's wrestling team and in spite of his size played guard on the football team.

Mike, who was popular among his teachers and the other kids, was usually in the middle of practical jokes and fun stuff, and when graduation was behind him, in the tradition of his family, he attended West Point.

In a class of 527 cadets, Mike Collins graduated 185th in 1952, modestly admitting his record was respectable but nothing to shout about, and he joined the Air Force to avoid accusations of nepotism had he joined the Army where his uncle was chief of staff.

Mike completed his flight training and moved on for advanced-day fighter training flying F-86 Sabres and learning how to deliver nuclear weapons all the while inching his way into test-pilot school.

While flying NATO duty in the summer of 1956, Collins was forced to eject from an F-86 after a fire started aft of the cockpit. He was safely rescued. Soon after he met Patricia Finnegan, his future wife. They had three children, two daughters and a son, and once Mike had accumulated over 1,500 hours of flying, he was assigned to the USAF Experimental Flight Test Pilot School at Edwards.

He was there when NASA named its third group of astronauts in June 1963. Mike found his name on the list and in three years he went into space with John Young on *Gemini 10* where they docked successfully with the Agena and he took a spacewalk and recovered a micrometeorite package.

But John Young and Mike Collins just didn't recover one micrometeorite package—they recovered two by finding and rendezvousing with the old *Gemini 8* Agena that Neil Armstrong and Dave Scott had to abandon.

All of it had been a whirlwind ride for Mike Collins who was a confirmed optimist.

"I'm not at all convinced that everything is going to work out well," Mike Collins would say, "but on the other hand, there's nothing wrong in thinking it should."

Like his crewmates, *Apollo 11*'s Lunar Module Pilot Edwin Eugene "Buzz" Aldrin Jr. was born in 1930. He was born a few months earlier than Neil and Mike on January 20, in Glen Ridge, New Jersey.

Buzz was the third child and only son of his mother Marion, whose maiden name was "Moon," and his father Gene Aldrin, a hard-to-please man who studied physics at Clark University under Dr. Robert Goddard, the father of American rocketry. In 1918, Buzz's father earned a master's degree at MIT in electrical engineering before becoming a pilot in the Army Air Corps, serving as an aide to General Billy Mitchell, who was regarded as the father of the Air Force and lobbied for the ability of bombers to sink battleships.

The overachieving father returned to MIT to earn a doctor of science (ScD) degree before becoming an executive with Standard Oil and battling with his son on whether or not he should attend Annapolis or West Point. Buzz won out and after turning down a full scholarship offer from MIT, he graduated third in his West Point class. His father simply asked, "Who finished first and second?"

The nickname "Buzz" originated in childhood: the younger of his two elder sisters mispronounced "brother" as "buzzer," and this was shortened to Buzz. It was also a term used by pilots when they "buzzed"—flew low over buildings and such to announce their arrival. The name set well with the aspiring young pilot, who in 1988 made it his legal first name.

Like his father, Buzz's ambition to excel had no bounds. Small for his age, young Aldrin picked a number of fights hoping to show he was fearless, and in neighborhood pickup football games, like a smaller Mike Collins, Buzz played with the older boys.

Following graduation from West Point, Buzz chose the Air Force and after earning his wings he fought in Korea—flying F-86 interceptors at the same time Neil Armstrong was flying his third round of Korean combat off the carrier *Essex*. By the time the fight ended in July 1953, Aldrin had flown 66 missions to Neil's 78.

But unlike Neil, Buzz did tangle with a few MIGs.

For his first kill, Aldrin said, "I simply flew up behind the enemy and shot him down." One of his gun cameras shot the first picture in the war of an enemy pilot bailing out, and Buzz's second encounter was more daring. With his F-86E he joined a faster formation of F-86Fs on an unauthorized hit of an enemy airfield inside Manchuria. Buzz shot down his second MiG some 200 miles from base and barely made it back.

Returning from the war, Buzz Aldrin pulled several assignments with one goal in mind—the Air Force's experimental test-pilot school at Edwards. He believed more education would help and he asked to be sent to MIT where, in three years, he completed a doctor of science degree. His thesis was "Line of Sight Guidance Techniques for Manned Orbital Rendezvous" and he became one of the fourteen astronauts in the third group announced October 17, 1963.

Neil did not know Buzz very well until they began playing musical chairs as the backup crew for *Apollo 8*, and the rumor mill sent out a story that other commanders could not work as well with Buzz as Armstrong could.

Neil simply did not find Aldrin a problem. He told me, "Buzz and I both flew in Korea. There was no question about his flying skills," he explained. "He was smarter than most. He liked to talk things through. He was a creative thinker, and he was willing to make suggestions.

"Besides, NASA needed a crew that knew what it was doing," Neil said flatly. "I didn't need beer-drinking buddies."

The crew Neil would take to the moon might have been a collection of misfits, but they were qualified, experienced misfits and flying with Armstrong each would know his job thoroughly.

While the flight crew operations experts were busy running simulations of the commander and the lunar module pilot getting in and out of the lander, and trying to figure out what made the most sense for the first step on the moon, Neil, Mike, and Buzz were busy training for tasks already decided.

As requested by Deke, Mike Collins spent his lion's share of time mastering the command module while Buzz spent equal time with the lunar module including knowing what was required to set up the lunar surface experiments.

The science work fell mostly to Aldrin with the job of the actual landing on the moon left to Neil. He was convinced the best training for the lunar landing was flying the Lunar Landing Training Vehicle and he never missed an opportunity to take the LLTV up.

"The trainer was harder to fly than the lunar module, more complicated," Neil said. "I had to land it in the wind and gusts and turbulence you don't have on the moon, and there was concern with mastering the times when we would undock the vehicles in lunar orbit. We had to be certain the computers would know the velocities and directions of each vehicle. If the computers got fooled or lost information in that process we could be in serious trouble.

"And keeping the command module and the lunar module safely docked was another worry," Neil explained. "There really were only three small hooks latching them together, and that's a lot of mass even in zero G for those small hooks to be handling. It could be a concern as well as the navigation systems

Neil and Buzz train in the lunar module simulator. (NASA)

knowing where you are at all times. We had to be sure we didn't drift out of outside limits. We may not be in a correct lunar orbit and we could crash into some higher-than-we-thought mountains on the backside of the moon."

NASA as a team had a sense of unshakable confidence but few could match the confidence of Deke Slayton and Alan Shepard. From day one of their long journey from Project Mercury to the planned lunar landings, they never faltered. Through successes and disasters, through triumphs and tragedies, through their own extremely disappointing groundings, they had only one goal. And for them there was no turning back. Together, with Deke as director of flight crew operations and Alan as chief astronaut, they selected the crews who would fly to the moon.

But NASA had yet to fly the very vehicle that would land America's astronauts there. It was a bug-eyed spidery creature called the lunar module—LM for short—that could only be flown in the vacuum of space, and had to be flown, successfully, on *Apollo 9*. That meant launching the entire Apollo assembly—the command module, the service module, and the lunar module—into Earth orbit and simulating as many lunar flight procedures as possible.

Once in orbit, *Apollo 9*'s commander and lunar module pilot would crawl into the LM, undock the two vessels, and test the LM's flying abilities before redocking with the command module. If it all went well, NASA would remain on schedule to land astronauts on the lunar surface before the 1960s were over if *Apollo 10* could then fly a demanding dress rehearsal to the moon and back.

If nothing went awry, if the two missions were successful, if the next great booster was ready, if the Apollo spaceships were ready to go, if no one broke a leg or came down with a bug at the last moment, if, if, if, then the first landing assignment would go to Neil Armstrong, Mike Collins, and Buzz Aldrin.

"*Apollo 11*'s launch was going to be a big moment no matter what our flight objectives were," Neil told me. "Until the lunar module flew we would not know if we could communicate with two vehicles simultaneously and separate at lunar distances. We didn't know whether the radar ranging would work. We didn't know a lot of things, and we knew too many things could go wrong on *Apollo 9* and *10*."

Mike Collins flatly didn't think they could avoid all ifs. He bet the Las Vegas odds would put *Apollo 11*'s chance of getting the choice plum at only one in ten, and the odds would improve to four in ten for *Apollo 12*'s Pete Conrad, Dick Gordon, and Alan Bean.

Nevertheless, the flight crew operations team completed their simulation runs on how the crew should leave the lander to walk on the moon—learning that with the inward-opening door of the lunar module it would be very difficult for the LM pilot to crawl over or scoot around the commander to get outside. The team decided that after the hatch was opened, it would be safe and prudent for the LM pilot to babysit the lunar module's systems until the flight director was absolutely certain all fluids and working parts in the lander were functioning as they should.

Then, if a problem should develop while the commander was on the surface, the LM pilot could begin readying the lunar module for liftoff and redocking while the commander grabbed a contingency sample of moon dirt and quickly returned in his bulky spacesuit to the open, unhindered area of the lander.

For the LM pilot to exit first would mean he would always be trying to maneuver around and over the commander. The flight crew operations team's decision was obvious: The commander should leave the lunar lander first and be the last to return. The method would be used for all following Apollo missions.

Before it was decided who should be first to leave the lunar lander, Neil told me he and Buzz talked about it with Aldrin suggesting they should get involved. Neil said he was aware of the historical significance, and added, "I just don't want to rule anything out—let's let the flight crew ops guys do their thing."

Unless Neil Armstrong, Deke Slayton, and Alan Shepard lied to me, Neil never lobbied to be first or offered an opinion. In his own words at one of those *Apollo 11* anniversary media briefings in response to the tired old question he said, "Whatever my crewmates might think, I had zero input, no input whatever into that decision," and to this reporter's knowledge, Buzz never officially challenged the decision.

And when it came to that lingering charge by some in the media that Buzz did not take a picture of Neil on the moon because he was angry, Neil stood up for Buzz. He told me, "I was the one with the camera. Buzz was busy every minute setting up the experiments and when he was done, near the end of our EVA, I handed Buzz the camera and he took some shots—one especially I like of me loading rocks on the LM."

Meanwhile another related question was making the rounds in official Washington.

It was reported the new president Richard Nixon was concerned that some overly ambitious Apollo commander, thinking this was his one and only shot at landing on the moon, might take unwise chances. Mr. Nixon, a space program supporter, asked the NASA administrator to tell Neil Armstrong if conditions became unsafe for the landing he was to abort, and the new president promised him he would get another mission—he would get another attempt to land on the moon.

Neil liked that, but knew he would never take chances with the lives of his crew.

I asked Neil if the story was true.

"Yep," he said, adding, "I was also told he made the same commitment to later crews."

When it came to training for the moon landings, Deke Slayton and Alan Shepard had one more message for their Apollo astronauts: "You people are going to live in the simulators and you're going to fly your own mission a couple of hundred times before you finally go out and launch." Deke and Alan demanded everything of them except corporal punishment and a few hundred push-ups. While one crew was training on the LLTV and on other exercises, another crew would be in the simulators. "You'll all follow the others' moves and listen to every detail of the debriefings. Ask all the questions you can muster." Deke continued, "Remember there is no dumb question and you'll keep learning from each other and like a growing snowball rolling down a mountain, gathering snow all the way, you'll be doing the same. You'll be gathering experience and information for your own flight. Remember the Russians are damn well not standing still.

"Our recon satellites have photographed one of their big N-1s on the launchpad," Deke told them, adding, "It's almost as big as the Saturn V, and if they can get the damn thing to fly, they can still beat us to a moon landing.

"We did great with *Apollo 8*," he assured them, "but we haven't won this thing by a long shot. Let's get cracking."

N-1 launches. The Russians are coming. (Russian Federal Space Agency)

SIXTEEN

———

REHEARSAL FINALS

A tall rocket gantry towered over the vast steppes of Kazakhstan, It was located on the out-of-the-way backside of Russia's sprawling Baikonur Cosmodrome. Despite its towering height it went unnoticed, hiding Russia's secrets. Only a window from space had a view thanks to an American Discoverer photoreconnaissance satellite staring down from its passing orbit. It blinked and took pictures of the rocket standing nearly as tall as the Saturn V.

Russian officialdom called their monster N-1. Famed Soviet rocket engineer

Sergei Korolyov, the father of the world's first artificial satellite Sputnik, worked in secret to prepare it for its job to boost history's first spacewalker, cosmonaut Alexei Leonov, to the moon—get him there and back before the astronauts landed.

But for Russia there was a setback.

America's Project Gemini was flying its final missions when Korolyov suddenly died in 1966. Unfortunately for the Soviets he left N-1's development without a firm hand on the tiller. Soon Russia's program to reach the moon was shredding itself. Without Korolyov, rockets were rushed to flight before they were ready. They exploded on the ground and in the air, and the Zond program to simply fly around the moon? It was abandoned in the wake of *Apollo 8*.

Korolev's brilliance had caused many to predict Russia would beat America to the lunar surface. They hadn't counted on the equal brilliance of Dr. Wernher von Braun and succeeding explorers in the White House. America's Saturn V had sent *Apollo 8* in orbit around the moon, and now, it appeared the only chance Russia had left was a successful N-1 flight.

The Russians took their one shot. The familiar launch ritual came to life. Expected countdown delays at Baikonur came and went, and on February 21, 1969, they secretly lit the monster. Engineers, technicians, launch controllers, and cosmonaut Leonov held their collective breaths.

The thirty rocket engines clustered together for N-1's first stage sent a river of rolling fire sweeping down the curving flame trenches, and Russia's biggest rocket ever blasted free. It heaved itself from Earth as the select assemblage chosen to watch cheered.

Alexei Leonov gripped the railing in front of him and shouted at the top of his voice. If N-1's launch were successful he had been selected to fly Russia's LK one-man moon "bug" to a lunar touchdown. Alexei Leonov would be the first human to step onto a place other than Earth.

But success wasn't to be.

What Leonov could not see was that as soon as N-1 was airborne its engines number 12 and 14 went "dark"—their fuel had been shut off by an internal computer that sensed something was wrong. Still with 28 engines running N-1 continued to accelerate, pushing its 34-story-tall structure into the area of maximum aerodynamic pressure. Right on schedule the remaining engines throttled back to reduce the shock waves of Max-Q. Then the monster was through the "shock barrier." At 60 seconds from liftoff the engines throttled back up to full power.

They shouldn't have. Instead of a smooth transition to its maximum energy, N-1's cluster of 28 remaining engines kicked to full bore trying to compensate for the loss of engines 12 and 14. The result was a tremendous vibration from trying to keep each individual rocket in sync. N-1's design was simply doomed. There was no way with 1969 computer technology to succeed in getting so many clustered rocket engines to work in sync and the effort shook N-1 violently. A liquid oxygen line came apart.

Fed by a shower of the best oxidizer known fires grew rapidly. Rockets overheated. Computers failed.

The flames spread faster and faster. Turbo-pumps tore themselves into blazing wreckage. Alexei Leonov was suddenly mute. He instantly knew he would not be going to the moon, knew he was witnessing the death knell of his country's lunar landing program.

Helpless, Leonov could only watch as high above, the growing flames grew at explosive speed as the escape tower attached to the LK unmanned spacecraft snatched it from the devouring fireball. "At least there was one thing good in this nightmare," Alexei quickly told himself as he witnessed a terrible conflagration of red flames replacing N-1 in the heavens, expanding instantly into a flowering rose filling the sky, burning all within its reach.

The blazing wreckage showered a waiting Earth. The white-hot debris left a footprint reaching 30 miles in all directions while in the stratosphere where N-1 had plowed into Max-Q pressures, flames billowed and grew, lofting upward in a mushroom cloud with a killer stem.

For Alexei Leonov it was obvious. It would now take a miracle for cosmonauts to reach the moon. He was overwhelmingly disappointed of course. But he was a devoted member of the family who dared to fly beyond the sky and he put his overwhelming disappointment in its appropriate box. From his heart he offered good wishes and good luck to his brother. "We ride with you, Neil Armstrong. Have a safe flight."

And Alexei would have been pleased to know when Neil was briefed on N-1's colossal failure, he, too, was saddened.

Ten days following N-1's collapse, *Apollo 9*'s veteran astronauts Jim McDivitt and Dave Scott along with rookie Rusty Schweickart rode their Saturn V rocket into Earth orbit. They would not be going to the moon. They would only circle Earth fully checking out the only major piece of Apollo hardware not yet tested in space.

But when Jim McDivitt first saw the lunar module he was astonished. "Holy

Moses, are we really going to fly that thing?" he asked, staring at the LM's aluminum foil outer skin. "If we're not careful, we could put a foot through it."

Five days after entering orbit and thoroughly testing and retesting the LM's parts and systems, McDivitt and Schweickart were feeling better about the ungainly spacecraft. The two astronauts opened hatches in the docking tunnel that linked their Apollo to the moon taxi and drifted weightlessly into the lunar module. They then sealed themselves off from Dave Scott who babysat the command ship while they orbit tested the LM's flying abilities.

For the next six hours they lowered their orbit, changed their plane, climbed back to their original orbit, and then flew a smooth return and docked with their Apollo command ship. There hadn't been the first hint of a showstopper.

The astronauts of *Apollo 9* fell madly in love with the ungainly flying machine that had been so carefully built for them by the Grumman team. It obviously had been put together with care and *Apollo 9*'s crew was happy to announce another mile marker to the moon had been crossed.

For the crew of *Apollo 11*, the training continued.

Neil stood clad in his bulky white spacesuit equipped with its backpack. Before him was a lunar module mockup and he was most pleased with *Apollo 9*'s success. That left only *Apollo 10* and Neil was acutely aware an accident like N-1's could still ground America if *Apollo 10*'s mission failed.

But he smiled. It was hidden within his helmet. He had aces going for his crew. If anyone could knock down that last remaining wall so he and Mike and Buzz could attempt a landing on the moon it was the *Apollo 10* crew.

Tom Stafford was simply the best. It was rumored he'd flown a cardboard box without wings. He had taught pilots to be test pilots and secretly had test-flown two versions of Russia's MIG fighters at the Air Force's secret base in Nevada known as Area 51. The U.S. had gotten the MIGs from Israel, and Neil wasn't about to forget that even though Stafford was destined to be a three-star Air Force general, he was a graduate with honors from the Naval Academy; and Cernan? What the hell! Gene was not only Navy and a graduate of Purdue, he'd been right there with him flying those demanding LLTVs. And John Young? Forget about it! He was the first of the Gemini Nine group to fly and John, too, was Navy. So what was left to be said?

Neil shook his arms and torso to better fit his suit and gear around his body as technicians made final adjustments to his 200 pounds of gear. He then walked stiffly and with effort across the space vehicle mockup facility to the lunar module trainer. There he stopped and caught his breath at one of the lander's bowl-shaped footpads. He then placed a gloved hand on the ladder leading up to the LM's crew cabin.

Every day *Apollo 11*'s astronauts were training to reach the moon, and this day he and Buzz Aldrin were training to leave the LM for their walks on the lunar landscape.

"Okay, you read me?" he asked.

"Read you five-square, Neil," answered the technician playing CapCom.

Neil then began moving through the EVA rehearsal with the technicians and training specialists while an already-suited Buzz waited to join him at the appointed time.

Neil struggled not only against the weight of his gear (it would weigh only one-sixth of Earth weight on the moon), but also against his suit's stiffness. Pressurized, he was working in a rigid exoskeleton making movement an effort. His gloves? Hell, they were like wearing pressurized baseball mitts. When it came to the camera it was almost impossible to keep it in his hands. That's why attaching it as part of his gear was a good move. He reached into a pocket on his suit's thigh and pulled out a collapsible long-handled scoop.

"Beginning the contingency sample," Neil said as Buzz waited and watched. The contingency sample would be his first duty during the EVA. He was to scoop up a sample of the moon right away so if he had to return quickly to the LM they'd at least have some lunar soil.

But the contingency sample was just in case. The plan was for him and Buzz to collect pounds of rocks and moon dust and set up experiments. All of it had to be rehearsed until their EVA duties were second nature. Now, following months of studies and restudies, planners had decided their outside activities would last about 2 hours and 40 minutes.

However long, it would be the highlight of their eight-day mission.

Two months later, following a successful launch and outbound flight, *Apollo 10* astronauts Tom Stafford, John Young, and Gene Cernan had their two ships—command module Charlie Brown and their lunar module Snoopy—docked and linked together approaching lunar orbit.

They were flying in the shadow of the moon—upside down in total darkness knowing they were curving around the lunar surface. They felt the moon

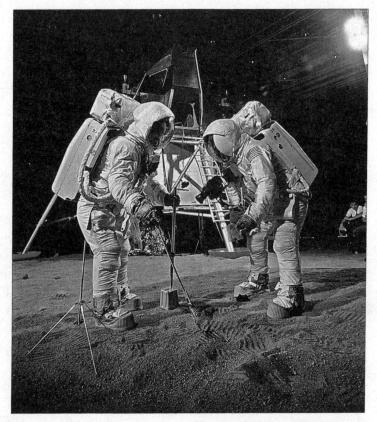

Buzz Aldrin joins Neil to practice their assignments on the moon. (NASA)

but couldn't see it. In fact they hadn't been able to see it all the way out, and then, finally, Gene Cernan caught a glimpse of the moon out his window. He watched as light bathed the lunar landscape and he called, "There it is! There it is!" and they were all suddenly startled.

They were finally seeing the moon they'd spent nearly three days climbing uphill from Earth to see—the moon John Young got into an argument with Mission Control about when he yelled, "We're not on the right damn course. There ain't no moon."

"Oh, you'll see it, John, just wait," CapCom assured him. They had all been feeling it, and hoping they wouldn't run into it, but just couldn't see it. Now there it was and they had their faces plastered to Charlie Brown's windows—

Apollo 10's crew flies out of the darkness for its first look at the moon. (NASA)

gawking, in awe. Before them were the lifeless jagged mountains, the wide deep craters, the smooth plains showered with boulders—cliffs soaring so tall and so close it appeared they could reach down and touch them. And in their awe they recognized they were seeing a world without life, a desolate dead place they had all been so eager to see. They feasted on this ghost world until suddenly Commander Tom Stafford reminded his crew it was time to fire Charlie Brown's big SPS rocket and enter lunar orbit.

Each returned to his station, and their linked spaceships eased into orbit around the moon's surface not only to further test the lunar module, but also to perfect navigating around the moon and to confirm a future landing site. The Sea of Tranquility, so named by ancient astronomers who thought it was a smooth sea of water, was *Apollo 10*'s main target. If one particularly level plain on that sea proved acceptable, then *Apollo 11* would be aiming for it. But for now it was up to Charlie Brown and Snoopy's astronauts to test-fly and scout.

When all was set Tom Stafford and Gene Cernan floated through the docking tunnel to enter and fire up Snoopy, while John Young stayed at his post as pilot of Charlie Brown.

Once undocked the two ships flew in formation for about 30 minutes with their astronauts becoming detailed observers. They were peering down through unfiltered sunlight at the craters and mountains, remembering how reporters dogged them with the question if they were going to ride Snoopy all the way down to 8.4 nautical miles, why not "go all the way?"

First, Snoopy was too heavy for its ascent rocket to lift it off the moon for a rerendezvous with its command ship, and equally important, there were still too many questions about the so-called "mascons," areas beneath the visible lunar surface. They were believed to be rocks of greater density that exerted higher gravitational forces. These perturbations of lunar gravity could cause dips in the Apollo command ship and its lunar module's orbits around the moon affecting the ships' navigational equipment. It would be the job of *Apollo 10* to measure this, to gather needed data to tie up the loose ends and bring them all together in a single tight package for Neil Armstrong and crew.

When it was time to separate John Young fired a burst from Charlie Brown's maneuvering thrusters and pulled away, while inside Snoopy Tom Stafford and Gene Cernan saw their ride home shrink into the distance.

"Have a good time while we're gone, babe," Cernan called, adding tongue-in-cheek, "And don't accept any TEI updates."

TEI stood for Trans Earth Insertion—the firing of Charlie Brown's big service propulsion rocket that would have the command ship leaving the moon for Earth. An update, a new TEI, would mean Young would be leaving without them.

"Don't you worry," John told him laughing.

It was the type of fighter pilot banter test pilots often used when their necks were on the line. The three were most aware Tom and Gene were flying a ship that could not get them home. If something went wrong, John would have to come get them.

The countdown to fire the descent engine that would send Snoopy barreling toward the lunar surface began when Stafford and Cernan were over the moon's far side. Mission Control went through another bout of nail-biting as the astronauts punched through critical maneuvers out of sight and out of touch with mission monitors.

Then they heard the excited voice of John Young from Charlie Brown; he'd appeared first around the limb of the moon, and as he reestablished radio contact with Earth, he fired off the initial message of mission progress: "They are down there," he confirmed, "among the rocks, rambling through the boulders."

Moments later Snoopy appeared, and the exuberant voice of Tom Stafford followed Young's report of boulder tripping. "There are enough boulders around here to fill up Galveston Bay. It's a fascinating sight. Okay, we're coming up over the landing site. There are plenty of holes there. The surface is actually very smooth, like a very wet clay—with the excavation of the big craters."

Cernan's voice, too, rang with unrestrained excitement. "We're right there! We're right over it!" he cried as Snoopy whipped moonward to within the planned 8.4 nautical miles of the Sea of Tranquility. "I'm telling you, we are low, we are close, babe!"

Stafford was suddenly back. "All you have to do is put your tail wheel down and we're there!" Snoopy swooped low over the moon, actually four miles south of the intended *Apollo 11* landing site because of the navigational errors planners had expected. Had this been the real thing here at the orbit's low point, Snoopy's descent engine would have been reignited for a final descent to the moon's surface.

But this time Tom Stafford and Gene Cernan would simply stay where they were, soaring to a height of 215 miles, their orbit's peak, before swooping down once more to 8.4 nautical where it would be time for the critical dismembering of Snoopy—separating the lunar craft so the legless upper portion would return them to Charlie Brown.

Astronauts Gene Cernan and Tom Stafford at the controls of the lunar module Snoopy, 8.4 miles above the moon. (NASA)

The two astronauts settled their lunar module into the needed attitude and flight for the separation. The larger descent stage and the smaller ascent stage were held together by four bolts that were to be blown apart by small explosives after which the ascent stage with its crew cabin would rocket away to find Charlie Brown some 300 miles ahead.

They were all buttoned up and ready to jettison Snoopy's descent stage when Stafford saw a yaw rate gyro indicating an intermitting failure. He immediately began troubleshooting.

The lunar module was equipped with two guidance systems. The primary called the Pings was used for flying the lunar lander down to and from the moon for rerendezvous. The abort system called the Ags should always be shadowing the Pings so if a problem called for the use of the Ags, the crew could hit the Ags anytime down to and up from the moon to rejoin the command module.

As they had gone through their checklist for the separation of the two stages, Gene Cernan told me, "The plan was to test the abort guidance system to make sure it worked, and I reached over and switched navigational control from Pings to Ags."

They had now set up Snoopy to find Charlie Brown by testing the abort guidance system. Moments later, during his troubleshooting of the yaw rate gyro, Tom reached over and inadvertently hit the guidance switch, changing it back to Pings.

Thinking they were ready for separation they blew the bolts and hell broke loose. Snoopy wheeled around in wild gyrations in radar search of its mother ship. Its snub nose pitched up and instantly pitched down. It then yawed violently between left and right. As close as they were to the lunar landscape the violent moves were terrifying, on the thin edge of lethal, and Cernan saw the surface corkscrew toward them, and he yelled, "Sonofabitch!"

The curse from Cernan sent instant alarm through Mission Control, but before controllers could react to what could have been a moment of danger Tom Stafford immediately killed the abort guidance system switch and took control of the LM manually.

As he gripped the controls, Tom realized Snoopy's thruster rockets had to stabilize the complete LM. He instantly jettisoned the descent stage 45 seconds early, getting rid of two-thirds of that weight. This gave him far less spacecraft to get under control and the veteran test pilot's skilled fingers went to work. Within 15 to 20 seconds the LM had calmed itself and settled into the desired attitude. Snoopy was ready to go find Charlie Brown.

"Tom, God bless him," Gene Cernan told me later. "He did a great job."

With their nerves and wits restored, Snoopy's crew fired its ascent rocket and charged ahead to find Charlie Brown. Tom and Gene were two of the best flying the lunar module and their docking with their command ship was *smoooooth*.

A delighted Tom Stafford told Mission Control, "Snoopy and Charlie Brown are hugging each other."

The three astronauts, back together in *Apollo 10*, made one more trip around the cratered landscape before beginning the journey homeward. John Young had a bit of information for schoolkids everywhere.

"About the man in the moon," he said, "We didn't see one here, but pretty soon there will be two."

Tom Stafford and Gene Cernan are happy to see their command ship Charlie Brown. (NASA)

John Young, at the controls of the command module Charlie Brown, watches as the lunar module Snoopy pulls up and parks. (NASA)

Back on Earth, the two who would be on the moon, Neil Armstrong and Buzz Aldrin, along with their command module master Mike Collins, moved into final training. Despite the odds makers in Las Vegas, the success of *Apollo 10* meant all but one of the "ifs" had been blown away. Only landing remained. *Apollo 11*'s crew couldn't wait to get its hands on what Tom, John, and Gene had learned.

In the coming days Neil, Mike, and Buzz grew exceedingly pleased with the superb job the *Apollo 10* astronauts had done. They had flown almost precisely the same track over the lunar landscape that *Apollo 11* needed, and had taken very detailed pictures of the descent and landing areas all the way down to the time of engine ignition, the time Neil and Buzz would begin their powered descent.

"*Ten*'s photographs were so useful we could commit to memory the major landmarks for our descent to the surface," Neil told me. "We could crosscheck

Snoopy moves in on Charlie Brown. Soon they will be hugging. (NASA)

every location. We should be able to make sure we were flying the planned track," he continued. "If we weren't, I could take over and I felt pretty good about flying Eagle to a safe landing site."

I could tell there was no longer a question in Neil's mind. The success of *Apollo 10* simply meant *Apollo 11* would be making the first attempt to land on the moon. The remaining question was when?

Deke Slayton huddled with NASA's top brass and then called Neil into his office. He stared at a man whose opinion he respected. He asked, "Are you ready?"

Neil was not a man to show overconfidence. But he couldn't help it this time. "Yes."

"That's your assessment?"

"You know you could always use more training, Deke, but we're there," Neil said flatly. "We should be ready for the July window."

Each month the Earth-Moon system rotated into position for the shortest flight between the two—a window of opportunity for the most desirable time

Tom, John, and Gene locked the route pioneered by *Apollo 8* into *Apollo 10*'s computers and headed home. (NASA)

to fly—and Deke Slayton gathered with NASA's top decision-makers again, telling them he had talked to Neil, and Neil said they'll be ready in July.

NASA announced on June 11, 1969, that the *Apollo 11* astronauts had received the go-ahead for a launch attempt July 16, with the first historic landing scheduled for Sunday afternoon, July 20.

Nine hundred miles north northeast of NASA's Manned Spacecraft Center, in Wapakoneta, Ohio, Viola Armstrong, Neil's mother, was doing housework when her radio reported, "Neil Armstrong and his *Apollo 11* crew has been given the go-ahead to attempt the first historic landing on the moon in July. The announcement was made . . ."

Tears suddenly filled Viola's eyes as she sank to her knees shutting out the radio's voice. She had long ago, even as a young teenager, given her life to Jesus Christ, and prayer for her was a daily practice. Suddenly hearing the announcement with her own ears, she was praying—praying as earnestly as she had ever prayed before that God would ride with her son, would protect those three young men all the way to the moon and back, and once she had fully asked the Lord for His protection and she was satisfied Neil's trip to the moon was God's will she climbed from her knees and called Steve, Neil's father.

She had never before been more proud.

While his mother went about reporting the news to family and friends, Neil was bracing himself to face the media.

NASA's policy was for each crew to hold a news conference approximately a month before launch and then sit for one-on-one interviews with the radio and television networks, the wire services, and the major papers like *The New York Times.*

Neil hated every moment of it. He simply wished reporters would let him direct his efforts at doing a safe and thorough job. The reporters who were Neil's friends knew the rules, and we lived by them. If you'd like his opinion the conversation was off the record, giving him the opportunity to consider your question freely. Then, if Neil told you something you considered news you asked permission to use it. Generally he gave his permission or gave you a reason why he couldn't.

NASA wanted to parade their astronauts before the public who was paying the bills, and reaching for the moon was a big bill. Neil knew he was riding on the taxpayer's dollar and, what the hell, he tightened his jaw and went along with the "dog and pony show."

First, Neil was asked, "Can you think of anything in which you aren't prepared?"

"The unexpected," he said confidently. "You've got to expect some things are going to go wrong, and we always need to prepare ourselves for handling the unexpected. We just hope those unexpected things aren't something that we can't cope with."

Another question was, "Will you take personal mementos to the moon?"

"If I had a choice, I would take more fuel," Neil answered with his one-of-a-kind grin.

"Will you get to keep a piece of the moon for yourselves?" another asked.

"At this time, no plans have been made."

The pattern of the news conference took shape in the form of quick questions with quick responses and Neil was feeling pretty good until he was cornered with *the* question: "Why should we spend the money to go to the moon?"

Neil shifted in his seat, and then looked directly at the reporter. "I think we're going to the moon," he said firmly, "because it's in the nature of the human being to face challenges. It's by the nature of his deep inner soul. We're required to do these things just as salmon swim upstream."

Neil and Buzz and Mike made it through the media's gauntlet and soon training was over and *Apollo 11*'s astronauts were packing lots of underwear. They had a trip to make with NASA furnishing flight coveralls and million-dollar spacesuits with special boots and gloves. Underwear and socks were just about all they needed along with a toothbrush or two.

Before leaving for the moon, Neil took care of more important things, such as wiping tears and caring for a cut on son Mark's small finger. (NASA)

SEVENTEEN

THE LAUNCH

Apollo 11's astronauts spent July fourth with their families before flying to the Cape. They were already in their 21-day, prelaunch quarantine to make sure they weren't exposed to some showstopping germ when their bosses strolled in wearing hospital masks.

The man with the most experience in spaceflight, Chris Kraft, asked, "Have we missed anything, Neil?"

"Nothing, Chris," Neil answered. "It's all been done. All that's left is the countdown." Kraft appreciated the confidence, and he agreed with Neil. If there was anything that hadn't been done, not a member of the launch team could say what it was. Kraft knew the equipment was ready. He knew the ground crews, the flight controllers—yes, the astronauts were . . . *Apollo 11* was the most ready

The *Apollo 11*-Saturn V stack loomed awesomely before sunrise. (NASA)

mission ever. He just wanted to make sure Neil didn't know something he didn't.

Then Apollo Director George Low had a question. "When you step off the ladder have you thought about what you're going to say, Neil?"

Neil took a measured, thoughtful moment. He was being questioned by the big boss and tact was demanded. The truth was he did have something to say in mind. It was also true he had not made a final decision. Neil had run it by his brother Dean and a couple of others close to him, and he told the big boss, "Sure, George, I've been thinking about it," adding a smile and quickly changing the subject. "Please tell all the hands that touched *Apollo 11*, all who worked so hard and for long hours, we appreciate it. This is their launch. Tell them they'll be riding with us all the way."

Three days before heading to the moon Neil had only a glimpse of the crowd gathering on the beaches and roadways, congregating in any place that brought them within eyesight or earshot of America's spaceport. A site with a clear view of *Apollo 11*'s launchpad was premium. There wasn't a room for rent in central Florida. It had come down to private families renting sofas, cots, even hammocks to anyone who wanted or felt like they had to be there. Neil thought he understood but he felt it was a lot to endure to see them leave for the moon.

The crowd of a million that came to witness Neil, Mike, and Buzz head for the first-ever moon landing found it easier if they came in their own car, RV, or camper. They squatted on beaches, roadways, and water edges surrounding the moonport. (NASA)

On the morning of launch Neil and crew were told the crowd had swelled to more than a million. Some 1,000 police officers, sheriff deputies, state troopers, coast guard, and marine patrol were struggling to keep the masses orderly. These keepers of the law had an estimated 350,000 vehicles and boats moving on the roads and waterways. Helicopters ferried VIPs to reserved bleachers and choice locations. NASA had invited 20,000 in all including Vice President Spiro T. Agnew, half of congress, representatives from the largest and smallest countries; but it was such notables as movie stars Jimmy Stewart and Robert Redford, and aviation heroes Charles Lindbergh and Chuck Yeager, that brought Neil a warmth of feel-good.

There was no one he admired more than Lindbergh, and Neil was most appreciative the great aviator had come. He was also grateful for one particular boat choking the larger waterways that were filled with celebrities' yachts and large cabin cruisers. On board one of the most luxurious vessels owned by North American Aviation, builders of the Apollo command and service modules, were his wife Janet and their sons Rick and Mark. Neil knew they were being well cared for as were his sister June and family, and his brother Dean and family, who were under NASA escort and secure within Florida's largest recorded gathering.

In Wapakoneta, the streets were virtually empty. Its 6,700 residents sat before their televisions, and at 912 Neil Armstrong Drive, the home of Neil's parents Viola and Steve, television networks had replaced their black-and-white set with the largest color television available.

Neil was happy his parents had attended his *Gemini 8* launch. They had been here to witness *Apollo 10*'s liftoff, too. But with the size of the crowd for *Apollo 11*, he was pleased his parents were staying home. He would later learn a reporter counted overnight 233 cars driving by their house.

Despite the crowds and all the excitement, the morning for Neil, Mike, and Buzz was all about going to the moon. Their day had begun three-and-a-half hours before liftoff when suit techs had dressed them in their spacesuits and helmets. From that moment forward, *Apollo 11*'s astronauts would be breathing manufactured oxygen—no outside air—until they returned in eight days. And when Neil led them from their crew quarters to their transport van to take them to their launchpad, he felt he belonged. Somewhere deep inside Neil was a lifetime feeling that his destiny was to take a "Lindbergh" step in flight. At least he hoped it was.

On their ride to the pad Neil reminded himself it all could have unfolded differently. It could have been Alan Shepard, Gus Grissom, Pete Conrad, Jim Lovell, or some other astronauts. But it wasn't. It was him and Mike and Buzz and he knew they had to get the job done. He glanced across the van at his two crewmates. He sensed they, too, had the same knowing. Like all their assignments before, they had gotten their jobs done, and going to the moon wasn't about the crew anyway. It was about science. It was about advancing science, humans leav-

A smiling Neil Armstrong leads the way from their crew quarters to the launchpad. (NASA)

ing their planet, leaving their cradles, following their destiny to explore and settle new places beyond Earth, advancing knowledge. Neil was satisfied.

He felt good about how everything had come down and when their transfer van reached their launchpad, he walked toward the elevator stopping before the lift that would take them to level 34. That's where the spaceships they'd named Columbia and Eagle waited. Neil had promised his mother he would give thanks before they lifted off for the moon, and he did.

Even though their helmets quelled the background noise on the pad, there was almost an eerie silence for Neil's moment of reflection, and he quickly recognized the silence was because of the absence of voices. During those times he'd been here for training, the launch stand and the service towers had swarmed with activity, workers in every direction doing what they needed to. Now the human beehive had almost disappeared. Saturn V was fueled. Most of the launchpad crew was gone; only those needed to load the crew and lock everything down remained. Before he started moving again toward the elevator Neil turned to acknowledge the man walking by his side.

Apollo 11 atop its Saturn V awaits its crew. (NASA)

Deke Slayton had ridden in the van with the crew to the pad and there would be no good-byes.

"Watch your asses and have a good trip," he told them as he watched them enter the elevator.

Minutes later, the three space-suited figures appeared on the crosswalk to the white room encasing *Apollo 11* and Neil stopped to stare down, giving Deke a final wave and taking a final look around before entering what would essentially be his home for the next eight days and 500,000 miles.

Deke did not deny that he wished he was going had it not been for doctors and heart irregularities, but it wasn't to be. The boss was comforted knowing Neil and Buzz and Mike were going for all of them—all who loved the sweet science of flight—Russia's cosmonauts, too, and Deke gave a good-bye thumbs-up.

On this morning of July 16, 1969, veteran NASA launch commentator Jack King was also ready to go to the moon. His voice from launch control boomed from

Neil leads his crew across the 34-story-tall level to board *Apollo 11*. (NASA)

speakers placed across the sprawling 180,000-acre facility and into every NASA location, the White House, and the worldwide radio and television networks. When King spoke the world listened:

After a breakfast of orange juice, steaks, scrambled eggs, toast, and coffee, the astronauts boarded *Apollo 11* at 6:54 A.M. eastern time. Commander Neil Armstrong was the first aboard. He was followed by Mike Collins. Buzz Aldrin, the man who is sitting in the middle seat during liftoff, was the third aboard.

Three hundred running the countdown in Launch Control had everything moving on time. Hundreds more worked with them in Houston's Mission

Control while thousands were up for duty at tracking stations around the world, aboard tracking ships at sea and monitoring craft in the air.

Launch Control on guard for the slightest hint of a problem. (NASA)

This is Apollo/Saturn Launch Control. We are now less than 16 minutes away from the planned liftoff for the *Apollo 11* space vehicle. All still going well . . .

Inside *Apollo 11*'s command ship the astronauts worked through their checklists, now fully separated from the outside world. Neil rested his left hand on the abort handle. With one twist the escape tower rocket would ignite and snatch them away to safety. For Neil aborting wasn't a big concern. He felt they'd covered it all. He was the only member of the crew with a window until they ejected the escape tower during powered flight. Those who planned such things knew the commander needed a window to make sure they were headed for sky instead of ground.

The three of them were in great spirits. They were confident. The countdown was sailing smoothly down through arming the escape tower, and range safety remained "green all the way." Even boaters and small aircraft were staying clear of the launch zone. Launch Control tested the systems for power transfer to the Saturn V. The lunar module Eagle was now alive on its own internal power.

> This is Apollo/Saturn Launch Control. We've passed the 11-minute mark. All is still Go.

Ten minutes. Neil, Mike, and Buzz's command ship, Columbia, was now on its own power. The massive crowd was silent. Only Jack King's voice was to be heard and it was suddenly musical.

> We've passed the six-minute mark in our countdown for *Apollo 11*. Now 5 minutes, 52 seconds and counting, and we're on time at the present for our planned liftoff at 32 minutes past the hour.

The launch team armed the destruct system, and the access walkway leading to the astronauts swung back out of the way.

Again Neil Armstrong placed his gloved hand on the abort handle. With their walkway to the gantry gone, the crew's only way to safety was to ride the instant thrust of *Apollo 11*'s escape rocket.

> "T-minus three minutes ten seconds," Jack King reported. "*Apollo 11* is now on its automatic sequencer."

The countdown was in the control of computers, and King said, "We're Go. The target for the *Apollo 11* astronauts, the moon, will be 218,096 miles away at liftoff."

T-minus 50 seconds and Saturn V was on full internal power.

> Neil Armstrong just reported back. It's been a real smooth countdown.
> Our transfer is completed on internal power with the launch vehicle.
> All the second-stage tanks now pressurized.
> Fifteen seconds and counting. Astronauts report they feel good.
> T-minus nine seconds.
> Ignition sequence starts.

From atop his skyscraper of a rocket, Neil heard the enormous burst of ignition—thunder echoing throughout the Apollo nullifying his hearing as flames blasted downward. 28,000 gallons of water each second smashed into the curving flame buckets to absorb and cool the volcanic eruption of seven-and-a-half-millions-pounds of thrust.

From atop his skyscraper of a rocket, Neil hears and feels the burst of ignition. (NASA)

Once again Neil Armstrong locked his heels under his seat and held on.

The Saturn V's first stage was alive, but it was anchored to its launchpad by huge hold-down arms, chained to Earth until computers judged it ready to fly.

"Six, five, four,"

Neil heard their Saturn V howling, heard chunks and sheets and flakes of ice falling steadily from coatings formed by the super-cold oxidizers and propellants. *Apollo 11*'s rocket was ready to leave. Neil knew it wanted to go, but the computers were still saying wait, wait another three seconds, wait until we can be sure.

"Three, two, one, zero, all engines are running."

Then, Neil felt it.

The most powerful machine ever built by man was suddenly free. Its holddown arms released their gripped and Saturn V screamed get the hell out of my way.

"Liftoff, we have liftoff, 32 minutes past the hour, liftoff of *Apollo 11*. Tower cleared."

Neil felt the Earth shake. Felt his crew's 36-story-tall stack of rocket and spaceships claw itself out of Earth's gravity well.

Birds flew for safety, wildlife fled for shelter, and the mighty rocket's shock waves slammed into the chests of the million-plus, rattling their bones and fluttering their skin and clothes. They were forced to lean into the powering wave of oncoming energy as *Apollo 11*'s Saturn V created its own earthquake, bellowing primeval thunder. Neil could hear and feel it all even through his helmet and earphones. Seven-and-a-half million pounds of thrust was slamming into the ground, bouncing back to his ears, but he keyed his microphone anyway and from his din of tumultuous sound told CapCom Bruce McCandless, "Roger, clock. We got a roll program."

Mission Control heard the report. Knew the clock was running and *Apollo 11* was rolling onto its proper heading. On the personal side, Neil knew only a few miles south of their ride to orbit—on a boat on the Banana River estuary were Janet, Rick, and Mark.

They had heard Jack King on the NASA squawk box announce liftoff but they couldn't yet see the Saturn V. They could not yet hear. They could only hold their hearts in their throats until *Apollo 11* came into view, and suddenly it did and their smiles grew as everyone on the boat screamed and shouted and bounced up and down in celebration. Before her new color television in Wapakoneta Neil's mother saw the ignition, saw the liftoff, and squeezed her hands in prayer.

But none of them knew the ponderous slow-motion rough and rocky ride

Neil felt the launchpad shake as the rocket clawed itself out of Earth's gravity well. (NASA)

Neil was feeling. The roar had been overwhelming but it was beginning to fade. He could now hear the slamming and banging and sloshing of millions of gallons of fuel and Neil reported to Mission Control, "Roll's complete and the pitch is programmed," adding, "One Bravo."

Neil, Mike, and Buzz were on their desired flight path. One Bravo was now their abort mode. They were high enough and moving fast enough to leave most of the noise behind and Neil felt he could now hear Mission Control despite the herky, jerky thunderous ride. It seemed all of Saturn V's stages were vibrating simultaneously as they flew through feathery white ice-crystal clouds—growing in weight. Their G load was building and they were slamming into the area of maximum aerodynamic pressure that would try to rip the Saturn V/*Apollo 11* apart.

Those in Launch Control take to the windows. (NASA)

But Max-Q's uncomfortable shaking would only be momentary, while back on the Cape the record-setting crowd saw a fire river eight hundred feet long trailing the *Apollo 11* train. They watched it pass the large flag standing before them with a ghostly ring of contrail dancing around the joints of Saturn V's stages.

They were at an altitude of 12 miles moving about 2,800 miles per hour, and CapCom told them, "Stand by for mode one Charlie."

Mode one Charlie meant flight controllers were checking the status of the upcoming first-stage burnout and its separation from the second stage.

"One Charlie," Neil confirmed.

"This is Houston, you are *Go* for staging."

"Inboard cutoff," Neil reported.

"We confirm inboard cutoff."

Neil, Mike, and Buzz braced for the "train wreck." Their G loads had them weighing four times what they did at launch and their first stage's five big rocket engines had compressed the Saturn's three stages like an accordion.

Apollo 11 waves good-bye to the flag. (NASA)

The moment those engines stopped burning, *Apollo 11*'s train lost its push and the astronauts were thrown back and forth against their straps. Neil wasn't at all sure they hadn't been thrown into the command module's bulkhead as he

Apollo 11's Saturn V leaving on flames two-and-a-half times its length. (NASA)

told CapCom, "Staging." The astronauts then heard metallic bangs and a mixture of clunks and clangs as explosive bolts blew away the now empty stage.

They were 40 miles high and 60 miles downrange, climbing faster than 6,000 miles per hour, and they heard more bangs and clangs from the second stage below as ullage rockets fired to settle the propellants. Then, the second stage lit off, kicking the astronauts back in their seats with renewed acceleration, and Neil sensed immediately the flight was much smoother and quieter—no more vibration with all sound left in their wake.

"*11*, Houston. Thrust is Go, all engines. You're looking good."

"Roger. You're loud and clear, Houston," Neil told CapCom, quickly adding, "We've got skirt SEP."

Apollo 11 losing one-half of its rocket train. We see the first stage burning out as the second stage ignites to push Neil and his crew toward Earth orbit. (NASA)

"Roger, we confirm skirt SEP."

The engine skirt was gone and a new sound slammed through the crew cabin. The escape tower's rocket had ignited automatically snatching away the no-longer-needed tower and the protective shield, uncovering their windows for the first time.

"Houston, be advised the visual is Go today," Neil reported.

"This is Houston, Roger. Out," said CapCom.

"Yes, we finally have windows to look out."

Neil was aware he was being ignored. The pleasure of having the ability to see beyond their spacecraft even though they could only see the black sky of space was not a priority for the busy team on the ground and he and Mike and Buzz were now enjoying the ride. Saturn V's upper stages had turned into gentle giants. They were now beyond the last particles of atmosphere and their ride was quiet and serene as smooth as glass and just a short distance away Neil suddenly saw tongues of flame lash briefly. Solid rockets on the Saturn V– discarded first stage were igniting to push away the stage from Apollo. No one wanted a "highway in the sky" collision at this point.

"Your guidance has converged; you're looking good," CapCom told Neil.

"Roger."

"*11*, Houston. You are Go at four minutes."

"Roger."

Apollo 11 parked in Earth orbit over a cloud-laden eastern Atlantic. (NASA)

Apollo 11 was now 190 miles downrange, 72 miles high, moving at 7,400 miles per hour and Neil told McCandless, "You sure sound clear down there, Bruce. Sounds like you're sitting in your living room."

Bruce came back. "You all are coming through beautifully, too."

"We're doing six minutes—starting to gimbal motors," said Neil.

"Roger, *11*," CapCom acknowledged. "You are Go from the ground at six minutes."

The second stage continued to burn and the *Apollo 11* train climbed faster and faster for another three minutes until it had emptied its tanks. Again the astronauts were snapped forward in their harnesses; again they were pushed back as the third stage lit off. They were moving along at 15,500 miles per hour, not quite Earth orbital speed, but the third stage would take care of that.

"Staging, and ignition," Neil told the ground.

"Ignition confirmed, thrust is Go, *11*. At ten minutes, you are Go."

Neil smiled. "Roger, *11* is Go."

Still heavily loaded with fuel needed to boost *Apollo 11* to the moon following an S4B (third-stage) burn to push them into Earth orbit *Apollo 11* was suddenly circling its home planet at a speed of 17,400 miles per hour.

This was the beginning of their planned holding orbit and Neil knew its purpose was twofold. First it gave the launch team a longer launch window and second it gave Mission Control and the crew two-and-a-half hours orbiting Earth to make sure astronaut and machine were ready to function far, far away.

They were entering orbit over the Canary Islands tracking station off the coast of Africa and Neil exchanged broad grins with Mike and Buzz. He released his harness and floated freely as if he were a feather with invisible wings. It was the feel-good feeling of weightlessness he had learned to love aboard *Gemini 8* and the treat alone made the trip worthwhile.

From Earth orbit, the astronauts' destination looms before them. (NASA)

EIGHTEEN

OUTBOUND

Apollo 11 sped around Earth for nearly two full orbits. Mission Control checked and rechecked its hardware. Flight controllers had to be absolutely sure the command ship and lunar module could carry the astronauts safely to the moon and back. Then CapCom told them, *"Apollo 11*, this is Houston. You are Go for TLI."

Neil Armstrong held up a gloved fist. "We thank you."

TLI (Trans Lunar Insertion) was the second major flight maneuver needed for *Apollo 11*'s crew to reach the lunar surface. Neil locked eyes with Mike and Buzz. They were smiling. No question they were ready. Each rechecked their seat harness, shifted and settled their weight, and waited.

"Apollo 11, this is Houston, *stand by!"*

"Roger," Neil acknowledged and the crew braced itself.

WHOMP!

"Ignition," he told Mission Control.

The powerful S4B third stage pinned them deep into their seats as every ounce of the big rocket's quarter-of-a-million pounds of thrust was burning. It was needed. They were reaching for the speed that would break them free of Earth's gravity. They were moving over the Pacific at 17,300 miles per hour. Their rocket would need to burn without stop, without even a hiccup for almost six minutes to reach 24,500 miles per hour—escape velocity.

They took comfort knowing what they needed was locked in their computers by *Apollo 8* and *Apollo 10*. Both previous flights were saying to *Apollo 11* follow us. They were, and Neil heard Houston telling him their Saturn V's third stage had been burning for one minute, "Trajectory and guidance looks good. The stage is good."

"Roger." Neil welcomed the news.

In the vacuum leading from Earth, the S4B's J-2 hydrogen fuel engine was burning a magnificent plume of pink and violet flame. Neil and crew felt its continuing power, felt its smooth and powerful ride. Their job was to monitor their instruments. CapCom told them, "Thrust is good. Everything's is good."

Neil returned a pleased "Roger."

They sat and rode and waited. They sensed they were leaving their home planet. Time seemed to have slowed and Neil wanted to pinch himself to be sure he was where he was—doing what he had so long dreamed of doing. Then there were only seconds to go in the long rocket burn, and CapCom was back: "*Apollo 11*, this is Houston. You're still looking good. Your predicted cutoff is right on the nominal."

Nominal—engineering lingo for normal, and Neil assured flight controllers, "Apollo *11*'s crew is Go." They were speeding away from Earth faster than a bullet and when they reached the end of their ride, they were moving across every single seven-mile stretch in only a second. The sudden loss of rocket power slid them into a free coast to the moon. With the exception of the hums of their Apollo's electronics and the fluids moving through its systems, they were left in a world of silence with little sensation of movement *and* the exhilarating freedom of floating weightless.

Speed: 24,500 mph
Earth Distance: 588 miles
Mission Elapsed Time: 2 hours, 54 minutes

Once again Neil could take a breath. He told Mission Control, "Hey, Houston, the Saturn V gave us a magnificent ride."

"Roger, *11*. We'll pass that on. And it certainly looks like you are well on your way."

"We have no complaints with any of the three stages on that ride," Neil assured them. "It was beautiful."

"Roger. We copy. No transients at staging of any significance."

"That's right," Neil told them. "It was all, all a good ride," he concluded as he and Mike Collins prepared for the next job: separating Apollo and the lunar module from the S4B stage.

"*Apollo 11*, you're Go for separation."

"Houston," Neil came back. "We're ready. We're about to Sep."

"This is Houston. We copy."

Neil felt explosive bolts fire, and heard metal clanging as he sensed his spacecraft was now smaller. "Sep complete, Houston," he reported.

They were leaving the final stage of their Saturn V behind and Mike Collins was now at the controls. The crew felt and heard Columbia's thruster rockets fire. Saw their ship turning around bringing them face to face with the Eagle. They were staring at their lunar module that had been riding atop the rocket secured inside its strong container. Columbia must have appeared as if it were a thick-bodied insect about to devour the helpless Eagle—but no matter. Mike Collins triggered more blasts from Columbia's thrusters and transparent flame streaked back. The tip of its docking probe entered the lunar module's docking port. One more blast of the thrusters and both ships rocked from impact. The astronauts felt it and heard metallic snaps as the docking hooks latched; suddenly staring back at them was the capture light. Their two spacecraft were now one.

Neil told the ground, "We are docked."

Speed: 19,180 mph
Earth Distance: 7,646 miles
Mission Elapsed Time: 3 hours, 29 minutes

For the first time since leaving Earth orbit and heading for the moon Neil could relax—only for a moment. "Houston," he began telling Mission Control, "you might be interested that out my left-hand window now, I can observe the entire continent of North America, Alaska, and over the Pole, down to the

Yucatán Peninsula, Cuba, northern part of South America, and then I run out of window."

"Roger, we copy."

Neil's view of a quickly disappearing Earth. (NASA)

Neil had to laugh out loud. He knew those in Mission Control had little time for sightseeing as they were getting ready for an evasive maneuver—one that would keep *Apollo 11* away from any contact again with its Saturn V's S4B third stage. The firing would slingshot the rocket past the trailing edge of the moon and into a solar orbit. But few—only the crews of *Apollo 8* and *Apollo 10*— ever had such a view of Earth. The small-town boy from Ohio was impressed.

"*11*, Houston," CapCom called. "Whenever you're possessed of a free moment there, we've got this maneuver PAD."

"Okay," Buzz answered.

CapCom read off all the letters and numbers needed for S4B's slingshot, and Buzz copied them while Neil and Mike began getting out of their sweaty spacesuits. They'd been wearing the heavy, pressurized garments since leaving their crew quarters six hours earlier. They quickly traded their exoskeletons for their comfortable, lightweight Teflon jumpsuits, but undressing and dressing in their newly acquired spacecraft was akin to performing the task in your family car.

As soon as Neil began feeling sorry for himself he remembered his friend and the mission's backup commander Jim Lovell who had actually lived two weeks in space in a phone booth—lived in one of *Gemini 7*'s two seats with Frank Borman. Neil quickly renewed his thanks for his spacious Apollo. It wasn't yet like living in a men's restroom.

Neil was most aware in weightlessness that for every action there's an opposite reaction. He would push against something and his body would take off in the opposite direction. Then he would have to muscle his way back all the while not really sensing how fast they were speeding away from Earth. When riding in a car, on a train, or in an airplane at night the moon and stars seem to stay in place—moving along with you as road signs and buildings and trees whiz by. These near objects tell you how fast you are going. In space you are too far away from other bodies of the universe to get a feeling of how fast you are moving.

Neil often pointed out that our Earth is moving around our sun at 67,062 miles per hour, and we pay little notice. We're only reminded by the change in the seasons.

Once he was comfortably in his jumpsuit, Neil was back at the window. The slowly changing panorama of Earth became a sphere and he was thankful he had kept his nose in the geography books in school.

He got a kick out of being able to pinpoint locations on Earth and he told Mission Control, "We didn't have much time, Houston, to talk to you about our views out the window. We had the entire northern part of the lighted hemisphere visible including North America, North Atlantic, and Europe and Northern Africa. We could see that the weather was good just about everywhere. There was one cyclonic depression in Northern Canada, in the Athabaska—probably east of Athabaska area."

Neil cleared his throat and continued, "Greenland was clear, and it appeared to be we were seeing just the icecap in Greenland. All North Atlantic was pretty good; and Europe and Northern Africa seemed to be clear. Most of the United States was clear. There was a low—looked like a front stretching from the center of the country up across north of the Great Lakes into Newfoundland."

Mission Control had been listening intently to the geography and weather report, and CapCom told Neil, "Roger. We copy."

Mission Control receives a weather report from the *Apollo 11* crew with this view of Earth. (NASA)

"I didn't have much to look at," Mike Collins added to the transmission, "but I sure did like it."

"We'll get you into the PTC soon and you can take turns looking," Mission Control told Mike.

PTC was the acronym for Passive Thermal Control. To make sure the astronauts and their ship weren't freezing on one side while roasting on the other they would slowly rotate *Apollo 11* to evenly absorb solar rays. This imitation of a rotisserie would give each astronaut equal opportunities to look out their window and admire the planet they were leaving and the lunar landscapes they were approaching. Flight operations had given the astronauts a simple tool in which to increase their view. It was a monocular—half of a set of binoculars—and with it the crew was taken with the fact that Earth appeared to be so fragile.

Neil thought of their home planet as small yet colorful, and compared to other space bodies it couldn't put up a very good defense against a celestial onslaught, yet it had been defending itself pretty well for 4.6 billion years. But there was that asteroid or meteor that took out the dinosaurs and the one that cleaned out the millions of acres of forest on the Siberian surface, and such an object if humankind doesn't prepare could take out Earth. That's what *Apollo 11*'s mission was all about and Neil was hopeful their mission would help life to continue and flourish.

There was a story going around that said when Neil was flying his F9F Panther on Korean combat patrol he flew over a ridge one early morning to see rows of North Korean soldiers sweating through their calisthenics. He could have mowed them down with his jet's machine guns, killing hundreds of them, but he didn't.

Neil never told me the story but it is an easy one for me to believe. Life simply meant too much to the man to waste it.

Once when we were having a drink at our favorite watering hole a young lizard scooted by my feet and I reached down with a cocktail napkin and picked him up. "Take him outside and let him go," Neil requested, and I did, releasing the young life by the pool to scurry away to live or to be eaten by a larger creature.

In Neil's mind, killing those defenseless North Korean soldiers would have been the same as shooting a helpless enemy fighter pilot parachuting from his burning aircraft. Neil Armstrong was on his way to the moon in hopes of pioneering new places for life to thrive beyond Earth. He sure as hell wasn't looking for ways to kill it. He could not and would not waste the smallest of God's gift.

Neil returned his attention back to the just completed S4B-stage slingshot maneuver and the crew of three agreed it was time to eat.

The crew downed sandwiches made from tube spreads of ham salad, chicken salad, and tuna, and for snacks they found their pantry stocked with peanut cubes, bacon bites, barbecue beef bites, and for the sweet tooth, caramel candy, dried apricots, peaches, and pears. The crew passed their thanks and kudos along to the chefs in Houston.

Speed: 8,161 mph
Earth Distance: 29,521 miles
Mission Elapsed Time: 5 hours, 55 minutes

The astronauts were spending their first day outbound grinning like kids in a down-home swimming hole. There was so much to experience, so much to do, so much to see watching Earth grow smaller in their wake as they took turns playing tourist guide with Buzz Aldrin describing the snow on the mountains in California for CapCom Charlie Duke. He added, "It looks like LA doesn't have much of a smog problem today, and with the monocular, I can discern a definite green cast to the San Fernando Valley."

"How's Baja California look, Buzz?"

"Well, it's got some clouds up and down it, and there's a pretty good circulation system a couple of hundred miles off the west coast of California," he told Charlie.

The astronauts sent television pictures to their worldwide audience and Mike Collins took the tour guide position.

"Okay, Houston, you suppose you could turn the Earth a little bit so we can get a little bit more than just water?"

"Roger, *11*," CapCom Charlie Duke replied with a grin. "I don't think we got much control over that. Looks like you'll have to settle for water."

"Okay, Charlie," Mike laughed, making room for Neil at the mike.

Mission Control had asked the crew for ten minutes more television, requesting a narrative from Neil describing what they were seeing.

Apollo 11 was about 60,000 miles from its homeport with the astronauts viewing a quarter of Earth. Neil began his ad-lib narration:

We're seeing the center of the Earth viewing a quarter of the sphere with the eastern Pacific Ocean. We have not been able to visually pick up the Hawaiian Island chain, but we can clearly see the western coast of North America. The United States, the San Joaquin Valley, the High Sierras, Baja California, and Mexico down as far as Acapulco, and the Yucatán Peninsula, and you can see on through Central America to the northern coast of South America, Venezuela, and Columbia.

Apollo 11 was about 60,000 miles from its home planet with this view of Earth. (NASA)

Mission Control thanked the crew for the television views as the fatigued astronauts realized it was time to call it a day.

A few preps later, the voice of Mission Control told the listening world:

This is Apollo Control at 11 hours, 29 minutes into the flight of *Apollo 11*. We don't expect to hear a great deal more from the crew tonight. At about 11 hours, 20 minutes we said good-night to them from Mission Control and they're beginning their sleep period about 2 hours early.

The early-to-bed was made possible because flight controllers found no need for firing *Apollo 11*'s thruster rockets for midcourse correction number two. The astronauts were currently on the desired course and any need to change their flight path would be handled in midcourse correction number three tomorrow. For now they busied themselves setting up their light mesh hammocks.

Two of the hammocks, much like sleeping bags, were stretched and anchored beneath the left and right seats while the center couch was folded down for the third crew member, the astronaut on call if needed.

Mike Collins took the first watch with a lap belt holding him in place in weightlessness. "It was a pleasure to doze off with no pressure points poking your body," Collins would later say. "What could be better than just floating all the way to the moon?"

Speed: 5,400 mph
Earth Distance: 63,850 miles
Mission Elapsed Time: 11 hours, 32 minutes

Mission Control and the flight surgeon kept a close watch on the crew reporting,

At 13 hours, 27 minutes into the flight of *Apollo 11*. Our flight surgeon reported a short while ago that command module pilot Mike Collins appeared to be sleeping soundly at this time. Biomedical data on the other two crewmen indicates that they are still awake.

After some 40 minutes, Mission Control was back:

The mission is progressing very smoothly. All spacecraft systems are functioning normally at this time, and the flight surgeon reports all three crewmen appear to be sleeping. For Commander Neil Armstrong, and lunar module pilot Buzz Aldrin, they appear to have begun sleeping about 5 minutes ago. Command module pilot Mike Collins has been asleep for about an additional 30 minutes to an hour.

The crew's sleep was uneventful for the most part with Mission Control reporting at 22 hours, 49 minutes ground elapse time. "The crew has been awake for some time according to the surgeon. Spacecraft communicator Bruce McCandless is standing by to make a call to the crew momentarily."

"*Apollo 11, Apollo 11*, this is Houston, over."

"Good morning, Houston. *Apollo 11*," Neil answered.

"Roger, *Apollo 11*. Good morning and when you're ready to copy, *11*, I've got a couple of small flight plan updates and the morning news."

The excitement of their first day was obviously the cause for the crew only sleeping five and a half hours, but in spite of their short sleep, once they were up they were raring to go.

Speed: 3,689 mph
Earth Distance: 114,204 miles
Mission Elapsed Time: 23 hours, 22 minutes

There were fewer chores on their second day leaving them with time to listen to music and marvel at the glow of Earth as they moved deeper and deeper into space.

Neil found his music soothing and a perfect fit for the occasion. In his teens he'd played the baritone, a large, valved brass instrument shaped like a trumpet, in a quartet called the Mississippi Moonshiners.

It was when Neil found chasing girls was fun. Playing the instrument he did he soon had the reputation of being the best kisser around. He carried this reputation to Purdue where he was a member of the university's concert band. At the moment he was listening to a recording of "Music Out of the Moon."

Neil's selection was very appropriate. What most on Earth could not know was that since leaving the planet *Apollo 11* had been in constant light with nothing to block the sun. It would be this way for the astronauts until they were almost to the moon.

Neil was really looking forward to their dark passage when he would have a clear, distinct view of the universe. That's when his backup Jim Lovell called, "*Apollo 11*, is the commander on board?"

Jim Lovell, first to open the very road to the moon they were now traveling, had been there every step of the way as Neil's backup. He was there ready to step in if for some reason Neil couldn't make the flight.

Armstrong rolled upright from his music-listening position, and quickly answered the call. "This is Neil, Jim, what's up?"

"I'm a little worried."

"Worried?" a puzzled *Apollo 11* commander questioned. "Worried about what?"

"You haven't given me the word yet," Lovell said. "You haven't told me to stand down. Are you Go?"

Halfway between Earth and the moon, Neil Armstrong is astonished by Jim Lovell's call. (NASA)

"Good Lord, Jim," Neil laughed. "We're halfway between Earth and the moon. "You've lost your chance to take this one, buddy."

"Okay," Lovell returned the laugh. "I concede."

Speed: 3,481 mph
Earth Distance: 123,307 miles
Mission Elapsed Time: 25 hours, 58 minutes

They sailed without the slightest bump across the halfway line in distance between Earth and the moon, but not yet across the gravity equalization point between the two bodies.

What Neil didn't know was that 23 years later the Galileo planetary craft would snap the first-ever picture of Earth and the moon showing the mission flown by *Apollo 11*. The aerospace editor of the Associated Press, Howard Benedict, one of the best damn friends Neil and I ever had, secured three copies of the photograph and I sent them to Neil. He was so impressed he signed a copy for this writer and one for Benedict to be treasured for a lifetime.

On December 16, 1992, eight days after Galileo returned for its second pass around Earth, the planetary spacecraft captured this remarkable view of the Earth-moon system from 3.9 million miles away. (NASA)

To complete their second day on their way to the lunar surface the feel-good astronauts put on an ad-lib television show with chef Mike Collins making a chicken stew, Buzz Aldrin doing push-ups, and Neil showing off by standing on his head.

They signed off their television show with a shot of distant Earth and then got to bed late. They slept for ten hours.

Speed: 2,406 mph
Earth Distance: 184,874 miles
Moon Distance: 73,732 miles
Mission Elapsed Time: 48 hours, 00 minutes

When they awoke, the astronauts took care of their duties including a midcourse correction before getting to the main event of the day: a meticulous checkout of their lunar module Eagle.

Mike opened Eagle's hatch and Neil squeezed through the two-and-a-half-foot-wide tunnel followed by Buzz.

Neil was looking for a scratch or any sign of damage while Buzz, the lunar module pilot, began preparing Eagle for its separation from Columbia some 45 hours later. Their inspection found nothing and they happily reported to Mission Control their lander was "immaculate" and ready to go.

With five-sixths of their flight completed, Earth's gravity diminished, and the moon's grip assumed dominance. Steady acceleration toward the small world required new thinking. The moon's mass was one-sixth of that of Earth's and it could be set down between the United States's Pacific and Atlantic coastlines. It was in all practical sense a dead world. On its surface the moon's horizon was much closer than on Earth and it was airless. No atmosphere or weather, and for their third sleep period, *Apollo 11*'s crew was restless. Neil, Mike, and Buzz knew what waited for them the next day. Entering lunar orbit was not a given. If their rockets did not slow their ship to the correct speed they would loop around the moon and return to the vicinity of Earth.

Speed: 2,823 mph
Moon Distance: 12,916 miles
Mission Elapsed Time: 71 hours, 31 minutes

When *Apollo 11*'s astronauts were awakened for their third and important day of entering lunar orbit, the moon's gravity was now pulling them faster and faster to their target.

"First off, it looks like it's going to be impossible to get away from the fact that you guys are dominating all the news back here on Earth," CapCom Bruce McCandless told them. "Even *Pravda* in Russia is headlining the mission and calls Neil the 'Czar of the ship.' West Germany has declared Monday to be 'Apollo Day.' Schoolchildren in Bavaria have been given the day off. BBC in London is considering a special radio alarm system to call people to their TV sets. And in Italy, Pope Paul VI has arranged for a special color TV circuit at his summer residence in order to watch you, even though Italian television is still black and white.

"Back here in Houston, your three wives and children got together for lunch yesterday at Buzz's house. And according to Pat it turned out to be a gabfest. The children swam and did some high-jumping over Buzz's bamboo pole."

McCandless went on to let the crew know the latest ball scores and told them, "Houston astrologer Ruby Graham says that all the signs are right for your trip to the moon. She says that Neil is clever, Mike has good judgment, and Buzz can work out intricate problems."

Neil grinned, quickly dismissing any thoughts of an astrologer as he and the crew turned to the important day ahead.

Mission Control was feeding *Apollo 11*'s astronauts the latest numbers and star settings for entering lunar orbit in 2 hours and 57 minutes.

Speed: 2,999 mph
Moon Distance: 8,430 miles
Mission Elapsed Time: 73 hours, 6 minutes

Apollo 11 was now flying through the moon's massive shadow and the view was testing the astronauts' nerves.

Here our nearest neighbor was looming before them, wrapping itself in a solar corona, while earthshine bathed its dark body with such illumination it was three-dimensional. The crew grabbed their cameras and a subdued Mike Collins asked Mission Control, "What sort of settings could you recommend for the solar corona? We've got the sun right behind the edge of the moon now."

"Roger."

"It's quite an eerie sight," said Buzz Aldrin. "There is a very marked sun's corona coming from behind the moon."

Solar corona of the moon as first seen by *Apollo 11*'s astronauts. (NASA)

"Roger."

"I guess what's giving it that three-dimensional effect is the earthshine," Buzz continued. "I can see Tycho fairly clearly. At least if I'm right side up, I believe it's Tycho, in moonshine, I mean, in earthshine. And of course, I can see the sky is lit all the way around the moon, even on the limb of where there's no earthshine or sunshine."

"*Apollo 11*, this is Houston. If you'd like to take some pictures," CapCom told them, "we recommend using magazine Uniform which is loaded with high-speed black-and-white film."

"And, Houston," Neil joined the conversation, "I'd suggest that along the ecliptic line we can see the corona light out to two lunar diameters from this location. The bright light only extends out about an eighth to a quarter of the lunar radius."

"Roger."

"Houston, it's been a real change for us," Neil continued. "Now we are able to see stars again and recognize constellations for the first time on the trip. It's . . . the sky is full of stars. Just like the nightside of Earth. But all the way here, we have only been able to see stars occasionally and perhaps through the monocular, but not recognize any star patterns."

Neil's view of a sky full of stars with Hydrogen Alpha located in constellation Orion 6,500 light-years distant. (*Ed Bianchina's Astronomy*)

"I guess it has turned into night up there, really, hasn't it?" CapCom asked.

"Really has," Neil agreed, trying to be factual and unemotional yet at the same time letting Mission Control know what they were seeing pretty much defied description.

They moved on out of the lunar eclipse and into a clear view of the moon they were approaching and Neil told flight controllers, "Houston, the view of the moon we have now is really spectacular. It fills about three-quarters of the hatch window, and of course we can see the entire circumference, even though part of it is in complete shadow and part of it's in earthshine. It's a view worth the price of the trip."

Apollo 11 approaches the moon. (NASA)

In Mission Control every monitoring console was in the green. *Apollo 11* was right on course with just ten minutes remaining before Neil, Mike, and Buzz were to fly behind the moon where their radios would be blocked. Cap-Com told them, "*11*, this is Houston. You are Go for LOI."

LOI (Lunar Orbit Insertion), the vital maneuver needed to reach the moon's surface, and *Apollo 11*'s crew was prepared for any contingency. The astronauts would be out of contact with Mission Control and just before they lost signal, CapCom radioed, "*Apollo 11*, this is Houston. All your systems are looking good going around the corner, and we'll see you on the other side, over."

"Roger," Neil assured them as *Apollo 11* vanished.

Speed: 5,225 mph
Moon Distance: 355 miles
Mission Elapsed Time: 75 hours, 26 minutes

Behind the moon it was as if *Apollo 11* and its crew of three didn't exist. They could not communicate with Mission Control. No telemetry or radio signals in, none out. The mission now existed only in their small world inside a spacecraft.

They were in their seats with their Apollo command and service modules docked with their lunar lander moving backward some 300 miles above the moon—a moon they would not see again until they turned around following the planned 6 minute, 2 second burn needed to brake their speed from some 5,000 miles to 3,000 miles per hour where lunar gravity would secure them in an orbit 61 by 169.2 miles.

The crucial rocket firing was set for 7 minutes and 45 seconds into their far-side pass and Neil and Mike and Buzz double-checked their settings—once, then again, and then a third time to make sure they did it right the first time. It had to be perfect. Just one digit in the computer out of place could send them into a lunar mountain or turn them and send them into an orbit around the sun.

But that wasn't going to happen. Neil and Mike and Buzz were sure of that. When they reached the mark, the three crew members felt the gentle ignition and heard the rumble of the SPS propulsion system burning. The numbers told them the rocket had fired and was running—burning smoothly and evenly for what seemed to Neil and Mike and Buzz an eternity. They stared into the blackness unable to see the moon—only sensing it—as their SPS propulsion system slowed their speed, moving them to within 61 miles of its surface. Their only worry was would the rocket burn too long, crashing them into the moon that was ever so near, and for 6 minutes and 2 seconds they rode *Apollo 11* to its slower speed. When it was over, finally over, it was a splendid and epochal moment: 75 hours and 55 minutes after ridding itself of its shackles on the launchpad, *Apollo 11* locked itself in lunar orbit.

No one on Earth knew that this had happened. In Mission Control this was a time of cliff-hanging suspense, a time to count the minutes and seconds that had to pass before *Apollo 11* emerged from the lunar farside to where it would hopefully signal success.

But on *Apollo 11* the celebration was already under way. The numbers were perfect. They had turned their spacecraft around and were looking down at the moon excitedly pointing out one spectacular feature after another and when they came around the lunar limb and Mission Control could hear them at the instant they should have, Earth celebrated.

"*Apollo 11* is getting its first view of the landing approach," Neil told Mission Control, recalling the pictures and maps brought back by *Apollo 8* and *Apollo 10.*

On one of the pictures Jim Lovell had taken from *Apollo 8*'s orbit only miles above the moon was a small lunar mountain. It was in the right spot for Neil and Buzz to ignite their descent stage. Lovell had named it Mount Marilyn. It would be *Apollo 11*'s landing marker.

"We're going over Mount Marilyn at the present time," reported Buzz.

"Roger. Thank you," acknowledged CapCom, quickly adding, "Jim Lovell is smiling."

Mission Control filled itself with laughter. Every flight controller knew the small mountain was named for a great wife and mother. Each also knew she had weakened only once—the day she had married Jim Lovell.

Neil's voice broke through the laughter. "Jim has given us a very good preview of what to look for here," he told Mission Control. "It looks very much like his pictures."

"This time we are going over the Taruntius crater, and the pictures and maps brought back by *Apollo 8* and *10* have given us a very good preview of what to look at here. It looks very much like the pictures, but like the difference between watching a real football game and watching it on TV. There's no substitute for actually being here."

"Roger. We concur," CapCom answered. "We surely wish we could see it firsthand."

"You will," Neil said. "There'll be lots more chances."

Snug in their lunar orbit, the *Apollo 11* astronauts had arrived and it was time to get the lunar module Eagle ready to land on the Sea of Tranquility. Neil and Buzz powered up their lander and moved through their list of checkouts before returning to Columbia for their fourth night.

Midday Sunday, July 20, 1969, Neil and Buzz were back aboard Eagle, and

From the command module Columbia astronaut Mike Collins watches lunar lander Eagle with Neil Armstrong and Buzz Aldrin move away for their landing on the moon's Sea of Tranquility. (NASA)

as they began their 13th orbit around the moon, Neil reported to Mike Collins, who was keeping the fires burning on board Columbia, "Eagle's systems are looking good."

"Hello, Eagle," Mission Control called. "We're standing by, over."

"Roger," Neil reported back. "Eagle is undocked."

"Roger. How does it look, Neil?" CapCom asked.

"The Eagle has wings," he told Mission Control, and once again billions watching on television held their collective breath.

From the Eagle, Neil and Buzz followed Mike and Columbia around the moon. (NASA)

NINETEEN

THE LANDING

Columbia came around the moon followed by Eagle, and Mission Control told them, "We're standing by, over."

"Houston, this is Columbia. How do you read me?"

"Roger, five by, Mike. How did it go, over?"

"Beautifully."

"Great. We're standing by for Eagle."

"Okay, he's right behind me."

Neil Armstrong and Buzz Aldrin were ready. They stood with their boots against Eagle's flight deck. They were flying backward with their faces parallel to the silent landscape below. Neil felt very comfortable with the astronaut in

Eagle in lunar orbit getting set to land on the moon. (NASA)

the CapCom chair in Mission Control. His name was Charlie Duke and he had already spotted a problem. Eagle's high-gain antenna needed tuning and Neil got right on it. But more important if a serious problem popped up, Neil knew Charlie would catch it right away.

"Houston, Eagle. How do you read?"

"Five by, Eagle," Charlie quickly responded. "We're standing by for your burn report, over."

"Roger, the burn was on time," Buzz told him. "It was perfect," he added as he began reading Charlie the numbers while Neil and flight controllers fussed with Eagle's high-gain antenna.

They spent the next several minutes reporting and confirming positions, measurements, and altitudes with Mission Control and then a final check with all involved, and CapCom told them, "Eagle, Houston, if you read, you're Go for powered descent."

Neil and Buzz hadn't read. Neil was still yawing Eagle's high-gain antenna trying to get it tuned just right, but Mike Collins above them in Columbia heard clearly. "Eagle," Mike called, "This is Columbia. They just gave you a Go for powered descent."

Neil and Buzz glanced at each other.

Powered descent. They were cleared to land.

They shot each other a wide grin and Neil calmly acknowledged Charlie with a "Roger."

Inside Mission Control's terraced rows of consoles and monitors manned by a small army of tense flight controllers sat a man who gave the finger to the long hair and outlandish garb that defined the 1960s.

With an outdated crew cut adding starkness to his features, Gene Kranz seemed strangely out of place. Yet he was the mission's final authority, the flight director whose sweeping powers would decide the what, when, and where of the first-ever descent to the moon's surface. Now, with Neil and Buzz dropping silently toward lunar dust and craters, no one called him by name. He only answered to "Flight."

He leaned into his communications panel and switched from the flight direc-tor's network to a flight-controllers'-only loop and told his team, "Hey, gang, we're really going to land on the moon today. No bullshit. We're really gonna do it."

Heads and bodies turned with smiles and thumbs-up to meet his words, and he switched back to the flight director's network so that the tense, nail-biting VIPs and families in the viewing rooms could again hear. Kranz, like the crew aboard Eagle, a former fighter pilot, appeared calm, but it was a front. Kranz's stomach was knotted. His heart thumped hard and loud, and when he lifted his right hand from the cover of the moon landing flight plan, he left behind a wet, perfect image of five perspiration-soaked fingers.

But more important than any other this day were the fateful words had been relayed across space: "Eagle, you are Go to ignite your descent engine." Neil locked his eyes on the glowing numbers displayed before him. He and Buzz were almost at the invisible junction of height, speed, range, and time when everything would come together. When the instruments would tell them they were 192 miles from their projected landing target, and precisely 50,174 radar-measured feet above the long shadows of the moon, they would unleash decelerating thrust and begin braking their speed for the touchdown.

Bright green digits changed constantly, the numbers flashing by in a breathless blur.

Then this was it. *PDI. Powered Descent Initiate.* On Earth, radio listeners and television viewers held their breath. Fingernails dug into palms.

Neil Armstrong preparing to land Eagle on the moon. (NASA)

Neil and Buzz braced themselves for ignition. At ten percent power there was no sudden hard smash of energy as Eagle's descent rocket was brought to life with a caress. Gently the ship descended through the black sky.

The Eagle's electronic brain monitored the deceleration, measured the loss of velocity, judged height, and confirmed the angle of descent. The invisible hand of the computer then began to add power. Gradually Eagle's descent engine increased to full power.

Flame gushed beneath them—glowing, gleaming plasma in a shock wave buoying in a vacuum and Eagle rocked from side to side. The lunar module's computer was so sensitive, alert, and instantly responsive, it fired control thrusters gently to hold the craft steady. Hollow thuds, distant subdued bangs, could be heard within the four-legged landing craft as small thrusters performed the ultimate balancing act.

Gravity pulled at Eagle far more gently than it would have on Earth as inside Neil and Buzz, who had been weightless, free from the weight of the heavy pressure suits and boots, were once again growing in weight. Their arms sagged. Legs settled within their suits. Their feet pressed downward in their boots as they yielded to their down-rushing speed.

Neil realized his eyes were tired, but he was alert with anticipation, immersed in the reality of the incredible adventure now before him, and he saw Buzz grinning like a kid.

Good Lord, they were going to land on the moon!

Fuel pumped through the lines under full throttle. Flame spewed far ahead and beneath them. The Eagle was in full fury now, blasting away weight and mass, slowing, slowing.

Headsets crackled at four minutes as Charlie Duke incredibly calm and professional called out, "You are Go. You are Go to continued powered descent. You are Go to continue powered descent."

"Roger."

But all was not well.

Back on Earth, Mission Control was thick with tension.

The highly trained flight controllers were focused on their monitors, tracking the curving line of the Eagle's landing descent. They were waiting for the vertical metal probe that extended from the landing legs of the ship to touch the moon's surface and signal a successful landing.

Those manning the front row of consoles were in what was known as the "trench." This was where final decisions were made, where ultimate responsibility lurked, where Deke Slayton placed his attention.

Deke had confidence in all those who worked in the supercharged atmosphere of the trench, but he was especially keen about a 26-year-old computer hotshot named Steve Bales. Like many in the control center he was young, but he was a seasoned veteran.

No one called him by his name. With a mission under way, he became GUIDO, the acronym for guidance officer. To the old-timers in the space business, Bales was pure genius. They referred to him sometimes as the "Whiz Kid."

Bales had been early for his shift, excited, filled with anticipation and wonder at what was coming. This was the most important, demanding, and exciting day of his life, and that sobering thought stayed with him as he took his seat at the guidance officer's console and nervously began twisting a lock of his hair.

Bales made a mental run through the list of possible signals that could sound danger alarms at any point in the epochal descent of the Eagle. And he knew that 24-year-old Jack Garman, in the back room, was running through the same mental checklist. Both were experts on the lunar lander's computers located deep within the bowels of Eagle.

These computer systems were essential to measuring all the electronic and mechanical forces and factors that would determine the success of Neil's and Buzz's touchdown on the moon. A landing soft enough to safeguard the health of the crew and maintain the structural integrity of the bug-eyed lander demanded a complex monitoring system. Changes in speed; rates of deceleration; shifting centers of gravity, weight, and balance; and engine thrust were all factors that were too sophisticated for the human mind to process. Computers—electronic brains—were designed to perform the superfast computations and to ride shotgun on everything that happened aboard the lunar module.

The computers aboard also contained sensitive electronic watchdogs. Alarm systems to detect imbalance, misalignment, and deviation from the exquisitely created flight plan. Only Bales and Garman were familiar with each of those alarms and what they meant. They were the only two people in that vast control system equipped to interpret any alarm emergencies aboard Eagle.

Everything they monitored within the landing craft was green and Go. The tension was there, but everyone was feeling pretty good about the descent.

Then, it all went out the window.

Suddenly Eagle's computers shrilled madly.

Alarm!

Emergency signals flashed within Neil's and Buzz's landing craft and one-and-a-half seconds later on consoles in Mission Control. No one expected a cry for help. Not now.

Eagle's descent engine had blazed at partial throttle for 26 seconds. Everything fit within the flight plan. But at 6,000 feet above the moon a yellow light flashed at the two astronauts.

On their way down to the Sea of Tranquility, Earth appeared in Buzz's window. (NASA)

Buzz had just told CapCom Charlie Duke, "Got the Earth right out our front window," when Neil quickly asked about the alarm: "Houston, you're looking at our DELTA-H?"

"That's affirmative."

Neil swore under his breath. "Program alarm!"

"It's looking good to us, over."

The commander of Eagle came back. "It's a 1202."

"1202," Buzz repeated Neil's reading.

Neil would have been less distracted by the computer alarms if he had known more about a simulation that had been conducted just a few days before their launch that indicated "executive overload" was not a problem, but even so he found the alarm a distraction.

What Neil relied on was the same conditions relied on by most test pilots—good velocities, good altitude, and the navigation was humming. There were no anomalies other than the computer saying, "Hey, I have a problem." There was nothing else complaining and his first judgment told him to keep flying, to believe his eyes, and he did, while Bales continued trying to prove the alarm was a condition of executive overload.

Bales and Garman knew for sure the warning was saying the computer was overloaded. So much was happening, so many performance signals were being generated that the computer could not absorb them all. What the two computer whiz kids didn't know was that Buzz Aldrin had decided to run both—not just one—but both of the systems Pings (primary guidance) and Ags (abort guidance) at the same time overloading the computer even more. The computer geniuses at MIT hadn't allowed for that—you can run one but not the other, but Buzz rightly judged both should be running should you need either. Neil, remembering how running both guidance systems at the same time almost sent *Apollo 10*'s Snoopy crashing into the moon, came back with a demand to settle the whole damn thing, "Give us a reading on the 1202 Program alarm, Houston, right now."

Steve Bales almost jumped out of his seat. All eyes were on him and he swallowed hard. He knew instantly the numbers he was reading were executive overload. He just didn't know why but instantly judged, "no harm." "*Go!*" he yelled. "*Go!*"

Surprised, Charlie Duke snapped, too. He didn't have time to wait. "Roger. We got— We're *Go* on that alarm."

Neil bit down hard. He'd buy that.

Let's go fly.

Thirteen hundred feet above the moon's surface, Eagle began its final descent. Flames gushed downward as the craft slowed. Neil had flown his mission right along the edge of the razor. He and Buzz functioned as one. Now they were

CapCom Charlie Duke, who would later walk and drive a lunar buggy on the moon, was on pins and needles during the landing. Backing up Charlie sitting on his left was famed *Apollo 13* commander Jim Lovell and lunar module pilot Fred Haise. (NASA)

doing more than falling moonward. They were so close Neil had to fly the lunar lander. He punched "Proceed" into his keyboard. The computer would handle the immediate descent tasks. Buzz would back up both man and electronic brain so Neil could adapt to flying in a vacuum.

A momentary smile crossed his face. Now let us see if those 61 LLTV training flights were worth it, the training he needed to land on the moon. Neil was betting they were. He was hopeful they had given him the skills to get the job done.

He looked through his triangular window and studied the desolate, crater-pocked surface before him. He had made many simulated runs, pored over dozens of photographs taken by *Apollo 10* marking the way, landmark by landmark, down to the Sea of Tranquility. He knew their intended landing site as well as he did familiar airfields back home, and he immediately noticed they weren't where they were supposed to be. Damn!

Eagle had overshot by four miles. A slight navigational error and a faster-than-intended descent speed accounted for their lunar module missing its planned touchdown spot. Neil studied the rugged surface rising toward him and Buzz noted a yawning crater wider than a football field. Eagle was running

Neil studied the rugged surface rising toward them. He had made so many simulated runs that he suddenly realized they had overshot their landing mark by four miles. (NASA)

out of fuel and headed straight for the gaping lunar pit filled with boulders larger than a Purdue jitney.

Scientifically it would be great to land next to and explore a crater gouged into lunar soil but Neil quickly ascertained the slope around it was too steep. If Eagle landed on a tilt they could never launch back into orbit.

With not a second to waste Neil realized he was on his own. This was where experience and training came into play and he looked beyond the crater. Landing Eagle was a matter of piloting skills he'd been honing. Against the wishes of Director of Flight Operations Chris Kraft, he had spent more time in the Bedstead, the Lunar Landing Training Vehicle, than any astronaut and now it was paying off. He needed to bring Eagle in to a smooth surface not by hovering and dropping, but by flying, by scooting across the lunar landscape as he had trained in the LLTV. There was only this one chance.

He gripped Eagle's maneuvering handle and translator in his gloved fists with a touch honed by years of flying the smallest and the largest, the slowest and the fastest—Neil knew the "thin edge" well, hell he had written it, and he had to fly as he'd never flown before. Knowledge, experience, touch—the skill of flying the *Gemini 8* emergency from orbit, bringing the X-15 rocket plane in from its Pasadena flyover, ejecting from his crippled jet fighter over Korea, and ejecting from the lunar landing trainer itself seconds before crashing—all of it, everything, came to this one moment.

Neil's fingers alternately tightened and eased on the maneuvering handle and translator as they sailed downward at 20 feet per second. He nudged the power, slowing to nine feet per second.

He attuned his senses to the rocking motions and the skids, sixteen small attitude thruster rockets kept Eagle aligned throughout its descent. A level touchdown was their ticket to safety, survival, and the return home.

Mission Control listened. They were mesmerized. They were in awe of the voices closing in on the lunar surface. Neil flew. Buzz watched the landing radar, called out the numbers that represented split-second judgment and flying skills.

Buzz was no novice. Jet-speed combat in his F-86 with Chinese fighters over the ugly mountains of Korea had brought him to this point. He had no questions about the pilot next to him. He was most aware of how Neil thought things through thoroughly and then did what he thought was right and he usually had arrived at the correct decision. Of all the pilots he had met and flown with, Buzz knew, without question none came close to Neil Armstrong. He was simply the best pilot Buzz had ever seen.

"700 feet, 21 down, 33 degrees," chanted Buzz.

"600 feet, down at 19.

"540 feet, down at—30.

"At 400 feet, down at 9."

"Eagle, looking great," Charlie Duke chimed in from Mission Control. "You're Go."

Despite the confidence of the astronauts' voices, there was still a problem: No place to land. Rocks, more boulders, surface debris strewn everywhere.

Neil fired Eagle's left bank of maneuvering thrusters. The larger rockets scooted the lunar module across rubble billions of years old. Beyond the eons of lunar debris, a smooth, flat area.

"On one minute, a half down," Buzz told him.

"70," Neil answered.

"Watch your shadow out there.

"50, down at two-and-a-half, 19 forward.

"Altitude velocity light.

"Three-and-a-half down, 220 feet, 13 forward."

"Eleven forward. Coming down nicely," Buzz told him.

Mission Control was dead silent. What the hell could they tell Neil Armstrong? Had they tried Deke Slayton would have killed them.

"200 feet, four-and-a-half down.

"Five-and-a-half down.

"120 feet.

"100 feet, three-and-a-half down, nine forward, five percent."

"Okay, 75 feet. There's looking good," Buzz told him as he stared at the obvious place Neil had chosen to land.

"60 seconds," Charlie Duke, told them.

Eagle had 60 seconds of fuel left in its tanks and no one wanted to think about it. If the descent engine gulped its last fuel before Eagle touched down, they would crash, falling to the surface without power.

What those in Mission Control did not know was that Neil wasn't all that concerned about fuel. He felt that once under 50 feet it didn't really matter. If the engine did quit, from that height at one-sixth the gravity, they would settle safely to the ground.

Neil calmly aimed for his new landing spot. He kept one thought uppermost in his mind: Fly. Eagle swayed gently from side to side as the thrusters responded.

Far away, in Mission Control, flight controllers were almost frantic with their inability to do anything more to help Neil and Buzz.

Deke Slayton knew they had to leave the landing to the pilots. But the clock was ticking away precious fuel. Charlie Duke looked at Deke and held up both hands, palms out. He didn't need to voice the question. Gene Kranz did it for him.

"CapCom, you'd better remind Neil there ain't no damn gas stations on that moon."

Charlie nodded and keyed his mike. A timer stared at him. "30 seconds."

"Light's on," Buzz told Neil as he watched an amber light blink the low-fuel signal.

Buzz then intoned the numbers like a priest, steady and clear, "30 feet, faint shadow."

"Forward drift?" Neil asked wanting to be sure he was moving toward known surface.

"Yes.

"Okay.

"Contact light."

Eagle's probe had touched lunar soil.

"Okay, engine stop.

"ACA out of Detent."

"Out of Detent," Neil confirmed. The engine throttle was out of notch and firmly in idle position.

"We copy you down, Eagle," Charlie Duke told them, and then waited.

Three seconds for the voices to rush back and forth, Earth to the moon and moon back to Earth.

Neil had to be certain. He studied the lights on the landing panel to be sure of what they'd just accomplished.

Four lights gleamed brightly—four marvelous lights welcoming them to another world where no human had ever been. Four lights banished all doubt. Four round landing pads at the end of the Eagle's legs rested, level, in lunar dust.

Neil Armstrong and Buzz Aldrin land on the moon. (Composite photograph, NASA)

Neil's voice was calm, confident, most of all clear, "Houston, Tranquility Base here. The Eagle has landed."

It was 4:17:42 P.M. EDT, Sunday, July 20th, 1969 (20:17:39 Greenwich Mean Time).

Charlie Duke spoke above the bedlam of cheering and applause in Mission Control.

"Roger, twainquility—Tranquility," a shaken and happy Charlie Duke answered. "We copy you on the ground. You got a bunch of guys about to turn blue. We're breathing again. Thanks a lot."

"Thank you." Neil permitted himself a grin even though he was doing his best to suppress whatever emotions he felt.

Pure, happy bedlam in Mission Control. (NASA)

In their excitement of the moment Eagle's crew simply shook hands. It was a defining moment in Neil and Buzz's lives—possibly in the historic significance of what had just happened.

"As the man who jumped off the top of the Empire State Building was heard to say as he passed each floor, 'So far, so good,'" Neil told Buzz, turning back to their checklists and their chores. They had no way of knowing how long it was going to take them to settle all of Eagle's fluids and systems, make their moon lander safe for its lunar stay.

To keep from worrying the public, NASA had hoodwinked the media by scheduling a four-hour rest and sleep period for the moon's sudden population. But the first two people inside a spacecraft on the lunar landscape would not be sleeping. They would be working feverishly to ensure they could stay long enough to take a stroll on the moon, and Neil looked at Buzz, "Okay, let's get going."

From 218,000 nautical miles Earth watches over Eagle on the moon. (NASA)

TWENTY

MOONWALK

Neil stared out at the alien world beyond his lunar lander's window. He was surprised at how quickly the dust, hurled away by the final thrust of Eagle's descent rocket, had settled back on the surface. Within the single blink of an eye the moon had reclaimed itself as if it had never been disturbed, and

Neil studied the desolation surrounding himself and Buzz. No birds. No wind. No clouds. A black sky instead of blue.

They had indeed landed on a dead world. A land that had never known the caress of seas, never felt life stirring in its soil, never felt the smallest leaf drift to its surface. No small creatures to scurry from rock to rock. Not a single blade of green. Not even the slightest whisper of a breeze. They were on a world where a thermonuclear fireball would sound no louder than a falling snowflake.

But there was no time for savoring it, or appreciating the science of it all. They had much to do very quickly, and they got busy. Surprisingly in only half the time they anticipated, Eagle had settled gently into its perch on the moon with all its systems purring. Neil and Buzz were ready to open the hatch. That plan to hoodwink the media with a scheduled four-hour sleep and rest period wasn't needed.

"Of course we wanted to get outside as soon as possible," Neil told me. "We needed the contingency sample to show we had been there, but we were convinced we'd need several hours to get Eagle's fluids and systems settled. With all that time passing and nothing happening," he explained further, "you reporters would have been speculating, guessing about possible problems, and we didn't want you guys inventing stories." Again that one-of-a-kind grin. "We wanted you thinking we were sleeping."

"Guilty," I acknowledged.

Neil and Buzz were ready to step onto the lunar landscape, and this reporter believed they were resting. The NBC News team was having dinner, celebrating, when we received the call, "They're coming out early."

With Texas beef and delicacies from Galveston Bay left on tables, we beat a path back to our microphones. We were in place to report the first human step on the moon when the last discernible bits of Eagle's atmosphere rushed pass its hatch's edges and we heard Neil tell an estimated billion plus listeners, "The hatch is coming open."

It was obvious NASA had made the correct decision regarding who would be first to leave the lunar lander. Outfitted in his bulky spacesuit, boots, and backpack there was no way Buzz could have maneuvered around Neil to the hatch. The commander simply had to be the first to leave and the last to return. Neil leaned forward, backing out, stopping on the porch with its large handrails leading to the ladder. Before he could begin descending to the moon's surface, he had to pull a D-ring, which lowered an equipment tray holding things needed for their moonwalk. It was called the MESA, and Neil told Mission Control, "The MESA came down all right."

"This is Houston. We copy, standing by for your TV."

The primitive, low-grade black-and-white television camera was located on the MESA, but the billion plus watching back on Earth didn't care about its quality. They wanted to see anything they could, and an excited CapCom Bruce McCandless told the astronauts, "Man, we're getting a picture."

"You got a good picture. Huh?" Buzz questioned.

"There's a great deal of contrast in it, and currently it's upside down on our monitor," CapCom explained. "But we can make out a fair amount of detail."

It was back to the days of fiddling with early television but Mission Control quickly readjusted the view and an excited McCandless reported, "Okay, Neil, we can see you coming down the ladder."

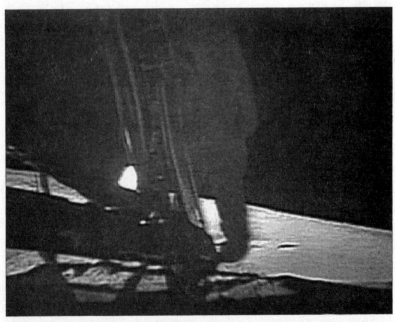

Viewers worldwide saw a strange, black-and-white image of Eagle's front leg with its ladder slanted across a totally dark sky. Below and in the background was a very bright lunar surface. On the ladder was a ghost. The ghost was Neil Armstrong. (NASA)

He moved slowly and steadily as if he had no place to go. The moon had been waiting for 4.6 billion years and Neil was in no hurry. Every move had to be precise, correct, no problems.

Soon he was a step above lunar dirt and he paused, staring at Eagle's landing footpads and legs. They had been designed to compress with the force of landing, making the ship more stable, bringing its ladder closer to the moon's surface. But Neil's piloting skills proved to be the problem. He sat Eagle on the moon so gently there was no collapsing of the pads and legs, and the bottom rung of the ladder was still three-and-a-half feet up.

Way to go, Armstrong, he scolded himself as he dangled a foot over the rung and fell slowly to the footpad beneath him. But before he would take another step, he wanted to be sure he could get back up to the ladder. In the low gravity he sprang with such force he almost missed the bottom rung. He steadied himself. Satisfied he could handle the extra long step, he descended back onto the footpad.

"Okay, I just checked getting back up to that first step, Buzz. It's not collapsed too far, but it's adequate to get back up."

"Roger. We copy," acknowledged CapCom.

"It takes a pretty good little jump," Neil told them before turning his attention to his dilemma. For some time he had been thinking about what he would say when he actually stepped on the moon. He had thought about one statement he judged had meaning and fit the historic occasion and he ran it by his brother Dean and others close. Neil had not made up his mind.

He told me he was undecided until he was faced with the moment.

He reached up with his gloved hand grasping the ladder, and then turned left, leaning outward. "I'm going to step off the LM now," he said, lifting his left boot over the footpad and setting it down in moon dust that shot up and outward in a fine spray—a spray that lasted only a quick instant in the absence of an atmosphere. "That's one small step for man," Neil said with a momentary pause, "One giant leap for mankind."

What most didn't know was that Neil had meant to say, "That's one small step for *a* man," and the loss of the "a" set off an argument for years to come. Had a *beep* in the transmission covered the a or some other loss of transmission wiped it from our ears, or had Neil nervously skipped the word?

Knowing Neil's struggles with public speaking, I believe the latter, and with all the excitement and emotions of the moment, I've never been convinced Neil himself really knew for sure.

His mother had told him her only real concern for his safety on the moon was that the lunar crust might not support him. Again Neil tested his weight. Then he told Mission Control, "The surface is fine and powdery. I can pick it up loosely with my toe. It does adhere in fine layers like powdered charcoal to the sole and sides of my boots. I only go in a small fraction of an inch, maybe an eighth of an inch, but I can see the footprints of my boots and the treads in the fine sandy particles."

"Neil, this is Houston. We're copying."

He stood there rock solid, boots braced for balance, enclosed in the elaborate pressurized exoskeleton that sustained his life in this inhospitable place. It was filled with energy, with supplies of heat and cooling, water, oxygen pressure—a capsule of life created by his Apollo colleagues, and Neil Armstrong stood looking long and hard at this small, untouched world.

He was overwhelmed; his sense and his thoughts set afire with the miracle of being on the lunar surface. He believed that he and Buzz and those who would follow were there for far more than just walking through lunar dust and measuring solar winds, magnetic fields, and radiation levels; all that was window dressing for their real purpose for coming.

It all condensed into every view they had of their fragile, beautiful Earth.

It was suddenly clear to this son of the land once walked by Orville and Wilbur Wright that he was on the moon to look back—to give every single human a clear look at spaceship Earth. In this neighborhood of the universe it was life's only world. It was encased in diamond-hard blackness and Neil recognized it mattered little if we were Republican, Democrat, Independent, apolitical, Christian, Jew, Muslin, Hindu, Buddhist, or whomever the hell we liked or disliked. We lived on a vulnerable world and we needed to take care of its very definite resources; on a world where we all would suffer terrifying consequences if we destroyed its ability to sustain us, its ability to foster and nurture the very life we now threatened to contaminate. Neil knew no matter how diligent, how great our effort to protect Earth, it was finite and one day if humans were to survive they would have to move on to new worlds. Helping to achieve that was what he and Buzz and all those who would follow were doing walking on the moon.

Neil stopped his thoughts, forced himself out of his introspection.

He and Buzz had much to do before they could catch a few hours rest and he turned and began walking farther away from the security of Eagle.

He knew with every step he was moving farther from the safety of his landing craft. The longer it would take him to get back the greater the risk, but with every halting step he was gaining confidence.

In one sense it was like learning to walk again—shuffling, stiff-legged yet buoyant, like wading through chin-deep water with his feet striking bottom— floating in low gravity within his spacesuit.

On Earth his exoskeleton weighed 348 pounds. Now on the moon it only weighed 58 pounds, and he told Mission Control, "There seems to be no difficulty in moving around, as we suspected. It's even perhaps easier than the simulations at ⅙G that we performed. It's actually no trouble to walk," he told them, adding, "The descent engine did not leave a crater of any size. It has about one foot clearance on the ground. We're essentially on a very level place here. I can see some evidence of rays emanating from the descent engine, but a very insignificant amount."

He had maneuvered Eagle to the flattest place on that part of the Sea of Tranquility. He pushed his boots into the soft, greyish-brown dirt. No living creature had ever done this before on this desolate, utterly silent world, but what puzzled Neil most was the lighting. From inside Eagle the moon's sky was black. But out on the surface it was still black, but a daylight black, and the surface looked tan. What he couldn't understand, if you looked down-sun, down along your own shadow, or into the sun, the moon was tan. If you looked cross-sun it was darker. If you looked straight down at the surface, particularly in the shadows, it looked very, very dark. When you held lunar material in your hands it was also dark, grey or black.

Neil's first task was to collect a contingency sample. If they had to abort the moonwalk early a small bag of lunar soil would make scientists happy. But he told himself he should do that in sunlight and for now he needed the camera. He needed to take pictures while his eyes were still adapted to the shadows.

"Okay, Buzz, we ready to bring down the camera?"

"I'm all ready," Buzz told him. "I think it's been all squared away and in good shape, but you'll have to play out all the LEC," Buzz instructed. "It looks like it's coming out nice and evenly."

Neil and Buzz used a special conveyor line with the acronym of LEC to lower the camera down to the surface, and Neil told Mission Control, "I'm standing directly in the shadow now looking up at Buzz in the window, and I can see everything quite clearly. The light is sufficiently bright backlighted into the front of the LM, that everything is very clearly visible."

Neil mounted the camera on a bracket on his chest and stepped forward to take the number one photograph. It was to have been his first footprint on the moon, but no sooner than he looked for it by the footpad than he was ready to kick himself. In his movements to check out Eagle's stance and operate the conveyor line to bring the camera down, he had walked over it. It was obvious his later steps had blotted out his first.

Then Bruce McCandless called, "We see you getting some pictures and the contingency sample, Neil."

Neil didn't move. He stood there disappointed with the loss of the first footprint, and McCandless asked again, "Neil, this is Houston. Did you copy about the contingency sample, over?"

No one was more aware than Neil how important the contingency sample was and he told Bruce, "Roger, I'm going to get to that just as soon as I finish these picture series."

Buzz watched as Neil completed the photographs and walked away to a sunlit area. He asked, "Going to get the contingency sample there, Neil?" "Right," Neil answered. "Okay. That's good," Buzz agreed.

Neil quickly reached into a thigh pocket and withdrew a collapsible handle with a bag on its end. He was in sunlight for the first time and he turned his back on the penetrating glare. He began digging into the surface. What he found surprised him. There was the same soft powder, but then there wasn't. He met resistance. "This is very interesting," he told Mission Control. "It's a very soft surface, but here and there where I plug with the contingency sample collector, I run into a very hard surface. It appears to be very cohesive material of the same sort." He scooped up enough lunar soil to fill the bag and told them, "I'll try to get a rock in here, just a couple." That'll give the geologists their money's worth.

"That looks beautiful from here, Neil," Buzz told him, talking about the sample, but Neil took Buzz's comment to mean the moon. "It has a beauty of its own. It's like much of the high desert of the United States. It's different but it's very pretty out here."

Neil got the contingency sample put away. Still holding the unneeded collector handle, he thought for a moment about throwing it like a javelin.

He smiled at how pleased Rick would be if his old man could set the record for throwing the javelin. How many times had he been on his older son's case to practice not only in Little League but in all sports, and he smiled, wondering for a moment what the record was, and then thought better of the idea. Instead he tossed it underhand and it sailed a long way, spinning in the sunlight.

Pointing down from Eagle was a 16mm movie camera loaded with color film. It was there to film the actions of the two astronauts on the moon and Buzz told Neil, "Okay. I have got the camera on at one frame a second."

"Okay."

"Are you ready for me to come out?" asked Buzz.

"Yes," Neil said and Buzz followed his instructions, asking, "How far are my feet from the edge?"

"You're right at the edge of the porch."

"Okay. Now I want to back up and partially close the hatch," Buzz reported, quickly adding, "making sure not to lock it on my way out."

"A very good thought," Neil agreed, as a wave of laughter rolled through Mission Control.

Buzz moved slowly down and then said "Okay, I'm on the top step and I can look down over the landing gear pads. It's a very simple matter to hop down from one step to the next."

"Yes, I found I could be very comfortable, and walking is also very comfortable," Neil agreed, pausing long enough to recheck Buzz's progress. "You've got three more steps and then a long one."

"Okay. I'm going to leave that one foot up there and both hands down to about the fourth rung up."

"There you go."

"Okay. Now I think I'll do the same."

"A little more," Neil suggested, "about another inch." There was a pause and then Neil shouted, "There, you got it!"

A second human was on the moon and Neil greeted Buzz at the bottom of the stairs as they again heard cheering in Mission Control.

"Beautiful view." Buzz grinned.

"Isn't that something?" Neil agreed.

"Magnificent desolation," Buzz said with feeling as he stared at a sky that was the darkest of black. No blue. No birds. No green below. There were many shades of grey on the surface and areas of utter black where rocks cast their shadows from an unfiltered sun, but no real color. Possibly tan under certain lighting.

Buzz Aldrin moves down Eagle's ladder to join Neil on the moon. (NASA)

The land curved gently but noticeably away—all the way out to the horizon that was only half the distance Buzz and Neil were used to seeing on Earth. But there on the moon, they could actually see they were standing on a sphere, and when they walked and looked down their motion fascinated them. Each time they took one of their half-walking and half-floating steps their boots set in motion a spray of lunar soil sailing outward and upward sharply and quickly without the hindrance of an atmosphere, and they even tried running and leaping strides that were impossible to do on Earth. But when they tried to sustain a jog, the mass and velocity created kinetic energy and stopping quickly was impossible.

It was as if they had found a new playground after school and they even

Buzz is firmly on the moon. (NASA)

The second human was on the moon and Neil greeted Buzz with the cheers of Mission Control. (NASA)

tried bunny hopping, an assortment of moves, and they wished they could stay on their new playground until they had explored every nook, every cranny; so much to see and do and so little time.

But despite their wish to drink in this new and strange and beautiful and wonderful place, Neil and Buzz had to move on to their chores.

First they reset their television camera's location 60 feet from Eagle. This would help Earthlings see some of the things they were seeing and it would let the world watch them go about their business.

Next on their list of duties was to plant the American flag. By international agreement no country could claim the moon, even the first to get there. That was stated firmly in a plaque on Eagle's front leg.

A plaque announcing Neil's and Buzz's arrival on the moon is secured to one of Eagle's legs. It is still there today for the next to come visiting. (NASA)

"For those who haven't read the plaque," Neil told the television viewers, "it says, 'Here Man from the planet Earth first set foot upon the Moon, July 1969 A.D. We came in peace for all mankind.' It has the crewmembers' signatures and the signature of the President of the United States."

Neil and Buzz then unfurled an American flag stiffened with wire so that it would appear to fly on the airless moon. But the moon's subsurface was so hard they could barely get its pole to stand. Once they did they moved back to clear the view of the 16mm color movie camera looking down from Eagle's window.

Then there were other protocols to meet.

Despite difficulty caused by the moon's hard subsurface, Neil and Buzz get the flag to stand. (NASA)

In the lunar dust, Neil and Buzz placed mementoes for the five-deceased American and Russian space flyers, Gus Grissom, Ed White, Roger Chaffee, Vladimir Komarov, and Yuri Gagarin (the first in space had died in a plane crash the year before), and one small cargo—private and honorable—carried by Neil. It was not to be divulged. It was a diamond-studded astronaut pin made especially for Deke Slayton by the three *Apollo 1* astronauts who planned to fly it on their mission before that dreadful fire. And there was one other remembrance. Very special and dear to Neil, a part of an unfinished life he so wanted to leave on the moon, and he did.

High overhead, aboard Columbia, Mike Collins continued orbiting the moon every two hours. "How's it going?" he called.

"I guess you're about the only person around that doesn't have TV coverage of the scene," CapCom told him.

"That's all right. I don't mind a bit. How's the quality of the TV?"

"Oh, it's beautiful, Mike. It really is. They've got the flag up now and you can see the stars and stripes on the lunar surface."

"Beautiful. Just beautiful," Mike Collins said, wishing he were there but knowing he had a much bigger job—keeping Neil and Buzz's ride home waiting and ready.

So the estimated billion people on Earth watching could keep track, Neil relocated their television camera to its cable's limits. Those viewing could now see even more and CapCom told him, "We've got a beautiful picture, Neil," adding, "Could we get both of you on the camera for a minute, please?"

"Say again, Houston," Neil asked.

"We'd like to get both of you in the field of view of the camera for a minute."

Buzz was in the middle of setting up experiments and while he found the request puzzling, he moved in front of the camera.

"Neil and Buzz," CapCom Bruce McCandless told them, "the President of the United States is in his office now and would like to say a few words to you, over."

"That would be an honor," Neil responded.

Buzz felt his heart increase its beats. He was surprised. Later Neil would tell him he had been told the president might call, but had not wanted to mention it until it was firm.

"Go ahead, Mr. President. This is Houston out."

"Hello, Neil and Buzz," President Richard Nixon began, "I'm talking to you by telephone from the Oval Room at the White House, and this certainly has to be the most historic telephone call ever made. I just can't tell you how proud we all are of what you are doing for every American. This has to be the proudest day of our lives. And for people all over the world, I am sure they, too, join with Americans in recognizing what an immense feat this is. Because of what you have done, the heavens have become a part of man's world. And as you talk to us from the Sea of Tranquility, it inspires us to redouble our efforts to bring peace and tranquility to Earth. For one priceless moment in the whole history of man, all the people on this Earth are truly one; one in their pride in what you have done, and one in our prayers that you will return safely to Earth."

There was a long silence, a grateful silence by a listening world, and Neil responded. "Thank you, Mr. President. It's a great honor and privilege for us to be here representing not only the United States but also men of peace of all nations, and with interest, and a curiosity, and a vision for the future. It's an honor for us to be able to participate here today."

"And thank you very much and I look forward," President Nixon told Neil, "all of us look forward to seeing you on the *Hornet* on Thursday."

"I look forward to that very much, sir," Buzz joined in.

What Neil and Buzz would find in the future when they were making their world tour was that they would be greeted by people everywhere saying "we," not just the United States, "We did it—we went to the moon!"

There was so much more they wanted to do but Buzz found the Sea of Tranquility more rugged than he'd expected. There were high and low areas—not the best place to set up the experiments but he managed to deploy a solar-powered seismometer to detect moonquakes and a laser reflector to help scientists measure the distance at any given time between Earth and the moon. Buzz and Neil were most pleased when Mission Control told them they were giving them an extra fifteen minutes.

Then when they had the solar-powered seismometer running, Tranquility Base appeared to be a fully operating scientific outpost. Neil left the experiments to Buzz and began moving about their landing site exploring on his own.

He quickly abandoned any thoughts of trying to reach and inspect that football field size crater he had to avoid during landing. But there was a smaller crater he'd flown over only about 200 feet away. He only thought about this one for a second and moved into quick strides to reach it. He found the smaller pit to be about 80 feet across and 15 to 20 feet deep. No larger than a good-size house.

While Neil took the pictures, Buzz set up *Apollo 11*'s experiments. (NASA)

A baby crater, Neil thought, "Muffie's Crater." He smiled, quietly remembering his and Janet's two-year-old who they'd lost to a brain tumor, and he permitted himself a moment. He stood there, remembering how Muffie would have loved sliding down into the pit. He had an overwhelming urge to do it for her. He'd love to have a sample of lunar bedrock anyway for the geologists. But then better judgment grabbed him. What if he couldn't get back up without the help of Buzz?

He settled for taking pictures and describing what he saw before heading back where Mission Control had put Buzz to work hammering a metal core tube sample into the hard subsurface. They then told Neil to gather rocks that would best represent their location—the Sea of Tranquility.

Once done, Buzz was pleased. He was leaving a scientific station on the moon. (NASA)

Muffie's Crater. (NASA)

With time running out they moved back to Eagle's ladder and Buzz was told to head back in. But before he did he took the camera from Neil and photographed the *Apollo 11* commander loading lunar material boxes on Eagle.

After completing all his duties Neil handed Buzz the camera. He had only a couple of minutes but he managed this one photograph of Neil busy loading moon rocks to bring back to Earth. (NASA)

Buzz handed the camera back to Neil and said, "Okay, adios, amigo."

Neil waved and watched him go up the ladder.

The two had packed their precious booty in sealed containers, and would now use their conveyor line to haul the boxes up to the ascent stage. Working with untried equipment in a vacuum, they struggled to get the samples aboard Eagle.

After a few starts and stops they managed to get it all on board and went through their long checklist, and then it was time to shut down history's first moonwalk on a place where a day lasts a month and time seems to crawl.

Neil sensed that if he came back to this same location on the moon a hundred, a thousand, a million years from now he would find the scene as he had left it. In his visit, he had little time to get to know this small corner of the solar system. Yet the knowledge and the samples from the moon he and Buzz were bringing back were priceless.

He joined his moonwalking partner inside Eagle to welcome the loud noise of oxygen filling their lander's cabin—the livable atmosphere they would need to take their helmets off. When they did they were met with a pungent odor— wet ashes and gunpowder. They were bringing the smell of the moon with them.

Overhead, Mike Collins and Columbia were streaking by and CapCom told him "The crew of Tranquility Base is back inside Eagle, repressurized. Everything went beautifully, over."

"Hallelujah," Mike yelled, and Mission Control reminded Neil and Buzz they needed to sleep for five hours before starting their countdown to rejoin Mike in lunar orbit.

The sleeping business was easier said than done. They were cold in Eagle. Whatever had been set up to keep them warm on this airless world wasn't doing the job, not forgetting they were wound up tighter than an alarm clock with accomplishment and excitement.

Neil Armstrong and Buzz Aldrin had landed on the moon Sunday, July 20, 1969, at 4:17:42 P.M. EDT.

Six hours and thirty-eight-minutes later Armstrong became the first human to set foot on the lunar surface at 10:56 P.M. EDT. Aldrin followed him 18 minutes later to become the second. Apollo's lunar landings would end after 12 Americans walked and rode in lunar cars across the moon's landscape. The last Apollo returned from the lunar surface December 17, 1972.

No human has visited the moon since.

Buzz says farewell to the moon and flag. (NASA)

TWENTY-ONE

THE RETURN

He was absolutely still, his weight next to nothing—almost floating, and he resisted awakening from such comfort, but he was cold—too cold to sleep. He thought about it—resting there in that wonderful consciousness of half-sleep and half-awake. He refused to open his eyes. He would just shiver and listen to the sounds that were trickling, whispering, humming along soft but persistent—an endless mechanical and electronic brook.

But ignoring it did not answer the question. Where in the hell was he?

Neil slowly opened his eyes and could gradually see his surroundings. He focused on the sounds of the bubbling brook—glowing circles, numbers, letters, buttons—and he saw Buzz standing at his window looking out at . . . at what?

The moon, Armstrong scolded himself. *You're on the moon. You were walking on it just hours ago and . . .*

"How was it sleeping on the moon, Neil?" Mission Control interrupted his waking thoughts.

"Cold." He spoke quietly, nodding for Buzz to answer for them.

"Cold," Buzz told CapCom.

They had not anticipated when they put the window shades in place and turned many of Eagle's systems off to sleep that they were turning off their source of heat. But despite the shivering, Neil had managed a couple of hours. Buzz told Mission Control, "Neil rigged himself a really good hammock with a waste tether and made him a bed on the ascent engine cover. He's just waking up."

"Great," Mission Control responded. "Tell 'im to grab a little breakfast, he has some flying to do," quickly adding, "We're pleased he got some sleep."

"Roger that," Neil heard Buzz answer as he stood up and stretched lightly in the low gravity.

The farm boy smiled—time to milk the cows, he told himself, coming fully awake. Until now they had been focused on reaching the moon, landing, taking a walk on its surface, setting up experiments, exploring, and gathering evidence. Now that they had done that and their "lunar booty" was on board, job one was to fly back to Earth and land near the aircraft carrier *Hornet* in the Pacific. Then, they'd meet the president.

But first they had to launch from this dead world and rejoin Columbia in lunar orbit. Mission Control had it all running smoothly with one exception. When climbing back into Eagle one of their backpacks brushed against the LM's ascent stage's arming switch and broke it. Buzz told Mission Control and the ground came up with a solution.

Neil and Buzz had dispensed with practically all tools in the interest of less weight, ensuring they'd have enough boost to reach Columbia. But they still had their space pens. They were told to retract the point of one of them and use the hollow end to flick the switch. It worked, and preparations were back under way.

In the command ship Mike had sweated every detail, every procedure in Eagle's prelaunch countdown. He recognized he was by being in lunar orbit in a safer place; not perched on another world depending on a single rocket to start him homeward. But the truth was he would have traded places with Neil or Buzz any time.

Columbia waits for Eagle to launch from the moon and join it again in lunar orbit. (NASA)

Now Eagle's ascent engine had to work. It had to burn long enough to reach some lunar orbit of meaningful height. If his crewmates could make it up to say 50,000 feet he could drop Columbia down and get them, but not much lower. There were mountains on the moon rising to 30,000 feet, and he didn't want to think about that knowing he would, if he had too, split a mountain to pick up his crewmates.

Mike did not know the thought of them being stranded on the moon forever had really never been considered by Neil. He knew if there was a problem they had been trained and retrained in several redundant ways to fire their ascent engine, and Neil believed the likelihood of that was zilch. Everything had been put in place to ensure their survival, but Neil's biggest concern was still the unexpected—had they failed to anticipate a showstopper?

One by one they moved through the countdown and when they had only ten minutes to go, Neil and Buzz stood side by side before Eagle's controls. They were set for the first launch from another place in the solar system.

Mike called from Columbia, "Neil, I'm reading you on VHF. You sound good."

"Yes, sir," Neil answered. "Couldn't be better, we're just purring along," and Buzz told him, "We're standing by at two minutes."

"Eagle, Houston," Mission Control joined the conversation with assurances. "You're looking good to us."

"Roger," Neil said, turning to Buzz. "At five seconds I'm going to hit Abort Stage *and* Engine Arm, and you are going to hit Proceed."

"Right," Buzz acknowledged.

Everyone knew if everything worked as planned, the astronauts were just along for the ride. At 5 days, 4 hours, 4 minutes, and 51 seconds elapsed time in their historic mission CapCom Ron Evans cleared Neil and Buzz for launch from the moon's Sea of Tranquility.

"Roger, understand," Buzz acknowledged. "We're number one on the runway."

"Nine, eight, seven, six, five," and Neil hit Abort Stage, Engine Arm, Ascent, and Buzz pushed the Proceed button. They waited—waited only for a moment. There was the muffled bang of pyrotechnic bolts firing and the ignition of their ascent rocket's hypergolic fuels, and they felt themselves lifting—a steady lift like a high-speed elevator, 21 hours after landing on the lunar landscape.

Eagle had blasted free of its bottom-half launch platform. Neil watched as flames from their ascent rocket ripped into the moon. Gold foil tore away from the landing stage, showering bits and pieces outward in all directions, and he saw the American flag they'd labored to raise whipped back and forth by their ascent engine's flames until it toppled to rest in the lunar soil.

Eagle's ascent rocket burned perfectly and Neil reported, "Climbing 26, 36 feet per second up," adding, "Be advised the pitch over is very smooth."

"Eagle, Houston. One minute and you're looking good."

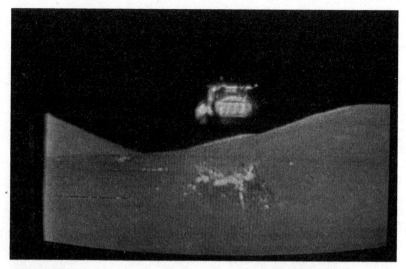

Apollo 11's liftoff from the moon was like this picture of *Apollo 17*. (NASA)

"Roger."

Beneath his feet Neil could see their landing site rapidly shrinking. If they'd had their television camera working by remote as *Apollo 17* later did, their viewers could have been watching their launch and climb. They could have seen Eagle race away on its ascent rocket's flames to become a pinpoint of light above the moon's cratered surface. But Neil knew their viewers could only hear him report, "A very quiet ride, just a little bit of slow wallowing back and forth—not very much thruster activity."

"Roger. Mighty fine," acknowledged CapCom Ron Evans.

Eagle was flying over the same *Apollo 10* landmarks Neil had been looking for during his landing approach to Tranquility. He told Mission Control, "We're going right down U.S. 1."

"Roger, Eagle. Four minutes. Everything's great."

Their ascent stage needed to burn seven minutes plus and Neil and Buzz stood with braced boots looking out their windows. The moon's landscape was growing smaller, but they had no trouble locating known craters along their track. "There's Ritter out there," Buzz said. "There it is, right there, there's Schmidt, man, and that's impressive looking, isn't it?" Neil nodded agreement watching one landmark after the other whiz by. He told CapCom, "Looking good here. It's a pretty spectacular ride."

Neil and Buzz could see the lunar landscape rapidly shrinking during their climb to rejoin Mike Collins in lunar orbit. (NASA)

"Eagle, Houston. You're still looking mighty fine."

Eagle's ascent rocket continued to burn, and then at the precise moment it completed its job with Buzz announcing, "Shutdown," before pausing and reporting the lunar orbit numbers: 54.3 miles by 10.9 miles. Mission Control cheered.

"Houston, the Eagle is back in orbit, having left Tranquility Base and leaving behind a . . . a replica from our *Apollo 11* patch and the olive branch," Neil said with a touch of relief.

"Eagle, we copy. The whole world is proud of you."

"We had a lot of help down there," Neil said with sincerity as he and Buzz settled in for the three-and-a-half hours they needed to reach their command ship and Mike Collins. The hard stuff was now behind them and they should easily reach Columbia, their ride home. All they had to do was fly the route pioneered by Snoopy with Tom Stafford and Gene Cernan two months earlier.

For nearly two trips around the moon Eagle chased Columbia. By flying a lower orbit taking less time to circle the lunar surface, the LM caught the command ship as planned. Eagle was about 40 miles behind and 15 miles below Columbia. Eagle could now be seen in the eyepiece of Mike Collins's sextant—a small dot moving up from the craters that made up two-thirds of the population of the moon. It was climbing steadily toward the other third and Mike would later say "It was the happiest sight of the whole mission."

At that very moment 27 degrees above the horizon, Neil made his move.

Mike Collins happily watched Eagle close the distance. (NASA)

Mike Collins watches as the Earth rises over the horizon behind Eagle. (NASA)

He fired Eagle's thrusters to complete the rendezvous, and told Collins, "We're burning."

"That-a-boy," Mike answered as he floated back to the command pilot's seat to watch the lunar module slowly close the distance, its thruster rockets spitting flame to cancel forward movement and complete its rendezvous.

"I see you don't have any landing gear," Mike told them.

They were returning with only half of Eagle, and with his movie camera rolling and his Hasselblad camera clicking, Mike captured the rendezvous on film.

Moments later, both ships moved back into radio contact with Mission Control and CapCom called immediately. "Eagle and Columbia, Houston is standing by."

Eagle coming home, with Earth watching. (NASA)

"Roger," Neil answered. "We're station-keeping."

"I've got the Earth coming up behind you," Mike shouted. "It's fantastic!"

They moved in for the docking and Mike watched Eagle drive steady as a rock down the centerline of final approach for the linkup. Minutes later the two ships were firmly docked—they were one once again.

Neil and Buzz floated back into the command module as all three astronauts celebrated their success and good luck knowing their flight had forever changed humankind's view of its place in this corner of the universe.

They were all full of laughter and happiness, and Neil and Buzz transferred their lunar samples into the command ship before discarding their faithful Eagle. They left it to orbit the moon until gravity tugged it down to its final resting place among lunar craters and plains that had been there for the eons.

When it came time to fire *Apollo 11*'s big service propulsion system to come home, what Mission Control called TEI (Trans Earth Insertion), CapCom Charlie Duke was back with them. Charlie didn't know he would go on to be the tenth astronaut to walk on the moon. He did know the firing was the next to last big concern for their flight. *Apollo 11*'s astronauts set up their main rocket engine to increase their speed to 6,188 miles per hour—far less than the speed they needed to break Earth's gravity. Then again when they fired it they would be on the lunar far side out of radio contact with Mission Control. As they disappeared behind the lunar limb, Charlie Duke in South Carolina vernacular told them, "Go sic 'em."

For twenty minutes Duke sweated the outcome. Then when *Apollo 11* reappeared Charlie asked, "*Apollo 11*. How did it go?"

"Time to open up the LRL (Lunar Receiving Laboratory) doors, Charlie," Mike Collins told him.

"Roger. We got you coming home," Duke assured them. "The LRL is well stocked."

Neil read all the burn times and numbers, and Charlie Duke said, "Copy, Neil. Sounds good to us and all your systems look real good. We'll keep you posted."

"Hey, Charlie boy, looking good here, too," said an elated Neil Armstrong loosening his demeanor a bit. "That was a beautiful burn. They don't come any finer."

Charlie just had to say it! "Only in Carolina," he laughed, and Neil and crew could not feel better. They were coming home. They had fulfilled President John Kennedy's call to land on the moon within a decade. Even more important, they had not become a permanent moon satellite. They were now down to only one real concern—penetrating Earth's atmosphere at the correct attitude to keep from becoming toast or skipping off the atmosphere to never return. They would need about two-and-a-half days to make the return trip, and the good news was the *Apollo 11* astronauts knew the way.

Apollos 8 and *10* had locked the flight path in *11*'s computers.

That first night going home proved to be the best night of sleep Neil, Mike, and Buzz had had during their mission. With a loud, fire station–style clangor to wake them if needed, the flight director permitted the whole crew to sleep for eight-and-a-half hours—until noon on Tuesday, Mission Control time. When the astronauts awoke they found their ship back in the grip of Earth's gravity, a point 44,620 miles from the moon and 200,100 miles from Earth.

Apollo 11 being welcomed by a half Earth and being bid farewell by a full moon. (Composite photograph, NASA)

The larger body was now pulling them back to family and friends, back to the life they loved. They took turns with the cameras at the windows.

CapCom told them, "The weather forecast for the landing area at the moment is 2,000 scattered, high scattered, 10 miles, wind from zero-eight-zero at 18 knots. You'll have about three-to-six-foot waves, and it looks like you'll be landing about 10 minutes before sunrise, over."

"Sounds good," Neil acknowledged, recalling another night landing on the Pacific; his and Dave Scott's hair-raising night return aboard *Gemini 8*.

And it was not only a night landing he wasn't crazy about. There was quarantine ahead—three weeks of it.

There simply was no evidence or scientific justification to quarantine the crew but Neil, Chris Kraft, Deke Slayton, and others who made Apollo go forced themselves to go along with the scientific community's fuss.

These hysterical handwringers were the same who had warned that a spacecraft landing on the moon would be gobbled up by hundreds of feet of dust and if NASA was to successfully place a human on the lunar surface, it would take the agency several attempts before astronauts could overcome the hazards.

These same doomsayers' forefathers had warned Columbus he would sail off the edge of the Earth, and now their descendants were warning that the *Apollo 11* astronauts would bring back alien organisms that would contaminate Earth. Their worries were so farfetched only Hollywood was capable of believing such nonsense, but that didn't stop some not-too-bright people in the media and on Capitol Hill from listening.

Following splashdown, *Apollo 11*'s crew was to be handed biological isolation garments, BIGs for short, that were rubberized, zippered, hooded, and visored with side-filtered face masks. After they struggled to put them on, they would

A smiling *Apollo 11* crew watched its home planet getting closer and closer. (NASA)

then be flown to the carrier where they would enter a sterile trailer. Upon reaching Hawaii, the quarantine trailer would be off-loaded for flight by a C-141 transport to the Manned Spacecraft Center. There they would be confined for three weeks, until doctors completed their examinations of crew, lunar rocks, and other samples, and until solid science confirmed for certain the astronauts would not unleash a plague on the world. Only then would *Apollo 11*'s astronauts be set free to tour the world. Everyone wanted to wish them well and shake their hands.

Neil thought quarantine was at best stupid, but politically mandatory. Having no choice, he and Mike and Buzz went along with a smile and they were pleasantly surprised when *Apollo 11*'s backup commander, Jim Lovell, called: "Just wanted to remind you that the most difficult part of your mission is going to be after recovery," Jim laughed.

"We're looking forward to all parts of it," Mike Collins answered with a straight face.

"Please don't sneeze."

"Keep the mice healthy," he agreed, quickly adding an up-to-date observation. "The Earth is really getting bigger up here and, of course, we see a crescent."

Early Thursday, July 24, the prime recovery ship the USS *Hornet* was ordered to move northwesterly a distance of some 200 miles.

It was said the *Hornet*'s new location would be in calmer seas and Mission Control told *Apollo 11*, "The *Hornet* will be on station just far enough off the target point to keep from getting hit. *Recovery 1*, or the chopper, is there; they're on station. And Hawaii Rescue 1 and 2, the C-130s, are within 40 minutes off your target point."

What CapCom failed to mention was the *Hornet* was moving well clear of any possibility of getting hit by a plummeting *Apollo 11* because President Nixon was on board. What CapCom also failed to mention was that President Nixon did not request the move, and he simply didn't know about it. High-ranking Navy officers who didn't wish to be responsible for the president getting injured or killed on their watch had given the order.

Then, once the crew was settled for reentry, Jim Lovell was back. He told them, "This is your friendly backup commander. Have a good trip, and please remember to come in BEF."

BEF meant blunt-end-forward. That would be with the heat shield leading, absorbing the heat, and Neil told Jim, "You better believe. Thank you kindly," and then turning he quickly added, "We can see the moon passing by the window, Jim, and it looks like what I consider to be the correct size."

"You're almost home, son," Lovell told him, adding, "You're going over the hill there shortly. You're looking mighty fine to us."

Neil simply said, "See you later," and with Mike Collins driving, and with *Apollo 11*'s command module moving about 25,000 miles per hour, they kissed the fringes of the atmosphere some 400,000 feet northeast of Australia. It was the beginning of their six-and-a-half-degree entry into Earth's protective cover. An entry too shallow would have them bounce off like a stone skipping across a pond never to return; too steep and their spacecraft couldn't stand the heat; but by entering just right *Apollo 11* was slowing and slowing, appearing to be a comet within a sheath of flaming colors.

The astronauts and their command module began a brief life as a man-made meteor. Temperatures on its heat shield soared to what could be found on the surface of the sun as they plunged downward with their backs to their line of flight. Only ships on the surface of the western Pacific were able to see

The USS *Hornet* standing by to recover *Apollo 11*. (U.S. Navy and NASA)

their command module hurtling through atmosphere, a flaming streak of intense and blinding fire with an outer red sheath followed by a streamer of flames 125 miles long. Inside the crew was cool and comfortable.

Apollo 11 was trading off tremendous speed for heat, and the more their heat shield burned, the slower the astronauts flew until they were out of the inferno of reentry and radio blackout. They heard, "*Apollo 11, Apollo 11* this is *Hornet, Hornet,* over."

"Hello, *Hornet,*" Neil answered. "This is *Apollo 11* read you loud and clear." They had left the blackness of the Pacific night and flown into the rising sun where their three large parachutes streamed away from the command module, opening partially for deceleration, then blossoming wide and full.

"What's your error of splashdown and condition of crew?"

Apollo 11 rode to Earth on three parachutes, like *Apollo 17* seen here. (U.S. Navy and NASA)

Neil, Mike, and Buzz dressed in their Biological Isolation Garments (BIGs) wait in the life raft attached to *Apollo 11* for helicopter pickup to the *Hornet*. (U.S. Navy and NASA)

"The condition of crew is good," Neil reported. "We're on main chutes and passing through 4,000 to 3,500 feet, on the way down."

A little more than a mile away the helicopter with rescue swimmers began its run toward *Apollo 11*'s splashdown point, telling the *Hornet*, "This is Swim One. We have a visual dead ahead about a mile."

"*Hornet*, Roger."

"This is Swim One, *Apollo 11*."

Neil answered, "300 feet."

The recovery helicopter replied, "Roger, you're looking real good," and the swimmers watched *Apollo 11* drop comfortably onto Pacific waves, and then reported, "Splashdown."

The first Earthlings to visit another place were home.

TWENTY-TWO

BACK HOME

Swim One's helicopter brought the *Apollo 11*'s astronauts to the aircraft carrier *Hornet*'s flight deck. A brass band was playing. Sailors and visitors, including President Richard Nixon were cheering. The crew, dressed in their greyish-green biological isolation garments (BIGs), was rushed through the back way into its sterile quarantine trailer.

The *Apollo 11* astronauts leave the *Hornet*'s helicopter in their BIGs. (U.S. Navy and NASA)

The three grateful-to-be-back astronauts immediately removed their BIGs and donned their NASA jumpsuits, then moved to the trailer's rear window. There stood Mr. Nixon on the other side.

Inside their sterile quarantine trailer, the astronauts donned their NASA jumpsuits for their meeting with the president. (NASA)

The president was grinning widely and through the glass his amplified words told them, "Neil, Buzz, and Mike, I want you to know that I think I'm the luckiest man in the world. I say this not only because I have the honor of being the president of the United States, but particularly because I have the privilege of speaking for so many in welcoming you back to Earth. . . . I could tell you about all the messages we have received in Washington," he said, and he did! He told them about the more than one hundred from foreign governments and their leaders who wanted them to come and visit. He told them about the millions of well-wishers. He continued saying just how proud everyone in America was of them, and then President Nixon smiled even wider.

I called, in my view, three of the greatest ladies and most courageous ladies in the world today, your wives. And from Jan and Joan and Pat, I bring their love and their congratulations. . . . I made a date with them. I invited them to dinner on the thirteenth of August, right after you come out of quarantine. It will be a state dinner held in Los Angeles. The governors of all the fifty states will be there, the ambassadors, others from around the world and in America. And they told me that you would come, too. And all I want to know—will you come? We want to honor you then.

"We'll do anything you say," Neil told the president, following with a little banter. Then Mr. Nixon's expression turned somber. "The eight days of *Apollo 11*," he said with deep sincerity, "was the greatest week in the history of the world since the Creation."

Moments later in the front yard of Neil and Janet's home she thanked hundreds of friends, proud neighbors, and the media who had been there for most of Neil's historic flight.

In Wapakoneta reporters wanted a statement from his parents. His mother Viola would only say, "Praise God from whom all blessings flow."

The president left the *Hornet* and the celebration subsided with the carrier getting under way for Honolulu. *Apollo 11*'s crew finally relaxed. The doctors drew blood and bodily fluids, and probed and kneaded muscles and flesh and bone, examining the moon visitors for any problems. When the medics were through Neil, Mike, and Buzz enjoyed some libations and great navy chow before collapsing into soft beds and real pillows. They slept solidly for nine hours.

Two days later they entered Pearl Harbor where thousands more cheered, more bands played, and more flags waved along with a broomstick on the *Hornet*'s mast signifying a clean sweep.

The worst part of Neil, Mike, and Buzz's brief Hawaiian stopover was going through customs. The best part was seeing their wives, even if it was only through the glass of their quarantine trailer.

The wives do Hawaii while the husbands wait. (NASA)

They were only in Pearl long enough for their quarantine trailer to be placed on a flatbed truck and driven slowly to nearby Hickam Field where their sterile unit was loaded into the cargo bay of a C-141 Starlifter transport for the long flight to Houston. Even though they were still confined to a small space, the crew was grateful for the additional room where they had a shower and hot food—even a cocktail hour with time to write down their thoughts and memories from their visit to the Sea of Tranquility.

They didn't reach Houston until after midnight but the crowds were still there. Thousands refused to go home. They cheered and screamed and waved at the trailer as it was placed on a flatbed truck again. And again there were more speeches and welcomes with the crew happiest to see their families.

Janet, Rick, and Mark spoke to Neil through the glass on a special phone hookup with his sons telling him how proud they were of their father, how glad they were to have him back. Mark proudly showed him his "all new" healed finger while Rick got in a little baseball talk. Their mother reminded the boys she had first dibs on their father—something about a "Honey Do List."

Sometime after 1:30 A.M., the astronauts' flatbed began to roll, but driving

Mike, Buzz, and Neil found their quarantine trailer nice, especially the comfort and libations. (NASA)

the truck with their sterile trailer wasn't all that easy. Even two hours after midnight along the way to the Manned Spacecraft Center people were still crowding the roadways. They were hollering and waving and trying to touch the astronauts' trailer. But the driver pushed on, made his way to the lunar receiving laboratory (LRL) where *Apollo 11*'s astronauts would spend 21 days in quarantine. That was the bad part. The good part was that inside they had a lab equipped with special air-conditioning, and a wide screen to view the latest Hollywood movies.

The lunar receiving laboratory was safe, secure, and its quiet was most appreciated by Neil. It had a population of three astronauts, two cooks, a NASA spokesperson, a doctor who was a lab specialist, and most important, a janitor.

The LRL was big enough for all. Neil kept in touch by phone, especially with Janet and the boys. Then there was his first call to his mother.

"Hello, Mom, this is Neil," he said with a feel-good voice.

"Oh, honey, how are you?" she cried with happiness.

"Oh, I'm just fine. All three of us are great—none of us got sick, Mom."

"Thank the Lord," she first said before asking her son about the moon. "You said it was pretty up there, honey, was it?"

"Yes, it was fantastically beautiful, Mom, but I got covered with dust. I can't get it off. I got it all over my nice white spacesuit. I can't brush it off—nothing will clean it, Mom."

"That's okay, honey, you bring it home," she told him. "Mom'll clean it for you."

Neil laughed softly. He loved putting his mom on. He couldn't tell her the Smithsonian already had what he wore for his walk on the lunar surface.

"I'm sorry Daddy isn't here, honey," she continued. "He just left for the farm. He'll be so sorry he missed your call."

"Tell him we're okay, Mom. Tell him I'll be seeing all of you pretty soon. And Mom," he spoke with tenderness, "take care of yourself and be careful. You're still my favorite girl. I love you."

Neil hung up happy and satisfied. He loved his family, the ones who nurtured him to full growth. Nothing relaxed him more than talking with his mom. Going to the moon was great, but it was also great to be back.

About the only time there was tension among the crew during quarantine was during a debriefing about seeing distant flashing lights in space. All three had

Apollo 11's crew in one of its many debriefings. (NASA)

seen the lights when they were outbound and their comments had crazed UFO buffs.

Now here in quarantine the conversation reached a fever pitch. You would have thought some were suggesting a UFO was following *Apollo 11* to the moon. Neil became annoyed.

"From the beginning I felt there was an explanation," he told me. "We were looking back at lights that steadily flashed—natural lights or human made it seemed to me, and we just didn't have the facts. I thought we had an obligation not to start some Hollywood frenzy about us being watched and followed by aliens."

Neil's analyses of the incident proved to be correct.

I had earlier learned a top-secret and sensitive American photographic reconnaissance satellite had failed and was tumbling out of control. With each tumble it sent out a reflection from sunrays giving the appearance of flashing lights. My source was solid and he only told me with the promise I would not report it. At the time it could have damaged the country's reconnaissance efforts severely.

The only UFO *Apollo 11*'s astronauts were seeing as Neil had suspected had been built here on Earth—one they could not identify, and when I told him, he laughed and said, "Well, what the hell, isn't that what a UFO is—an unidentified flying object?"

As always Neil was right on target.

If anyone was concerned about what happened to the lunar rocks and soil the *Apollo 11* crew brought back they needed only to be in the lunar receiving laboratory watching those standing outside a vacuum chamber dressed in hospital whites. They were geologists and one shoved his hands into a special, leak-proof set of arms attached on the other side of the wall in the vacuum chamber. He then placed his gloved hands on a silver box and opened the container. Inside were some of the moon rocks harvested by Neil and Buzz. They were still preserved in a 4.6-billion-year-old lunar vacuum and once removed amazed and startled geologists marveled at the charcoal-colored lumps and dust that one called, "burnt potatoes!" Now they were looking at a mystery.

It would be another three decades before computer models would tell them an infant Earth and moon were products of a solar system smashup. An incoming planetoid had gouged a great wound into our planet leaving it aflame in the hottest of fires and wracked with quakes.

A wounded Earth's gravity grabbed the planetoid and dragged the nearly destroyed space traveler into an orbit around its surface where it recollected and repaired its wounds to become the moon we see today.

Most of the heaviest elements from the planetoid, especially its iron, remained deep inside the now-molten Earth, beginning a long settling motion to the core of our infant world. The impact sped up Earth to a full rotation once every 24 hours.

The geologists in the lunar receiving laboratory had no idea that they were looking at scorched soil from the twins that created our Earth-Moon system.

What they would soon learn from the materials brought back by *Apollo 11* and the landings that followed was that Earth and the moon are much alike, and lunar-orbiting spacecraft mapping the moon would cast aside their long belief that our lunar neighbor was without water.

Near the moon's south pole is Aitken basin, one of the largest known impact craters. About 1,600 miles across—the equivalent of more than half the distance across the United States—the crater has some of the highest mountains in the solar system. Water has been detected there, but scientists believe water cannot persist on the moon's surface. Sunlight quickly vaporizes it so it can only be found in cold, permanently shadowed craters like those located in Aitken.

If this were correct, then obviously Aitken's shadowed lands with livable temperatures would be the perfect place for humans to establish a lunar colony with meteorite repelling domes. Their biggest problem would be keeping them filled with an Earth-like atmosphere.

The geologists studying *Apollo 11*'s moon rocks soon learned they had underestimated Neil Armstrong. He had honed his basic knowledge of science learned in classrooms and he was interested. He had studied their questions and had collected superb samples. In debriefings with the geologists he was full of detailed comments on what he had seen, and he made it clear a scientific observer should be part of future Apollo flights.

The one thing that would later disturb Neil about the lunar materials he and Buzz had so meticulously collected was that some moon rocks had been given to many heads of state for diplomatic reasons. In some cases leaders leaving office kept them or sold them to the highest bidder.

In this was a lesson. Neil needed a shield against the money grubbers who would be coming after him, enticing him to trade his celebrity for enrichment. That was not for Neil.

Even though he never had a fat bank account Neil never felt his family was deprived. He never felt they were poor. He was simply a person of science, a gatherer of knowledge. He left the glad-handing, the rubbing elbows, and the get-rich schemes to others.

His time in quarantine not only gave the doctors full access to *Apollo 11*'s astronauts, it gave Neil time to write letters, make notes, and think about his future.

He quickly realized he could only go to the moon once. There could never be another project or adventure that could top that achievement. The downside was it had come when his life was only half completed. What was he to do for the next forty years? Family and learning would play the biggest part. Teaching would play another. He had much to think about.

Neil's thoughts were interrupted by Deke coming in and sitting before the glass. The boss turned on their two-way communications and asked a question lingering in the *Apollo 11* astronauts' minds. Would you like to fly again?

Neil smiled. He was aware of the John Glenn thing.

After John became the first American to orbit Earth, President Kennedy told NASA a hero of Glenn's stature should not fly again. His life should not be put at risk. Neil, not one to put on airs, was pretty certain NASA would treat the first human to step onto the moon the same.

He saw clearly his value to the agency was to star in its dog and pony shows. He would preach the sermon—a sermon in which he believed while realizing he didn't care a whit about the adulation. He was equally certain he would be more comfortable on the moon than before an adoring crowd.

Quarantine ended August 10, and three days later Neil said, "This is the last thing we're prepared for," as he, Mike, Buzz, and families visited three cities in one day—New York, Chicago, and Los Angeles.

The three-cities-in-a-day victory lap was the beginning of a monthlong worldwide tour. President Nixon got them started by having Air Force 2 fly them and their families to New York City.

By all accounts not even Lindbergh's record-setting 1927 parade approached the size of the crowd to cheer the *Apollo 11* astronauts through a blizzard of ticker tape down New York's Canyon of Heroes between the skyscrapers. When the final count had been taken four million had celebrated *Apollo 11*'s achievement, including this writer.

Neil had only one complaint: "Those who tossed whole stacks of computer punch cards out of windows weren't aware some of the stacks didn't come apart and they hit like a brick. There were dents in our cars and bumps on our heads."

No New York City ticker-tape parade was greater than *Apollo 11*'s. (NASA)

As wild as the celebration had been in New York, the crowds were even wilder in Chicago. By the time they arrived in Los Angeles for the presidential state dinner in Beverly Hills, the astronauts were deaf from shouting, their smiles were frozen, and their fingers were crushed. None was looking forward to shaking another hand.

Even their formal wear was more comfortable than the noise of the crowds.

As promised, President Nixon and his wife Pat along with their grown daughters Julie and Tricia hosted the astronauts and their wives in their presidential suite. Joining them were former first lady Mamie Eisenhower, Esther Goddard, widow of America's father of rocketry Robert Goddard, and many government notables including governors from 44 states.

Governor Ronald Reagan of California was very interested in their flight, and they were most pleased with all the movers and shakers, the famous and the celebrities, who were there. They had never been invited let alone honored at a presidential state dinner.

Neil was especially happy to see Jimmy Doolittle who led the first bombing raid on Tokyo—B-25s taking off from aircraft carriers for just 30 seconds over the untouchable city only weeks after Pearl Harbor. Doolittle was also the man who headed NACA when Neil had been hired by the flight science agency in 1955.

Doolittle, Lindbergh, Gagarin, Shepard, and Glenn were Neil's heroes and there was another—Wernher von Braun, the great rocket scientist, the father of the Saturn V that had so flawlessly boosted them to the moon. He was there, and each member of the crew took special note of the presence of Jimmy Stewart, Bob Hope, and so many of their heroes from the movies.

Neil had hoped Charles Lindbergh would be there. The famed aviator had been invited but chose not to come out of his self-imposed seclusion, a decision Neil would soon appreciate. In years to come he was grateful he had the opportunity to meet Lindbergh on several occasions.

"I had enormous admiration for him as a pilot," Neil said. "I'd read some of his books. I was aware of the controversial position he took on certain issues. But I was very pleased to have had the chance to meet him, and I think his wife Anne Morrow Lindbergh was a wonderful person and quite an eloquent writer."

Following all the congratulations and speeches at the dinner Neil was most happy to see his own family. His parents as well as his grandmother and sister and brother and their families were there. He grabbed as much time with them as possible.

The three-city tour and presidential state dinner were only the kickoff. Next they threw ticker tape and confetti at them in Houston along with the ultimate Texas barbecue in the Astrodome. Frank Sinatra was master of ceremonies.

Neil then got in gear for the coming four weeks. He hit the worldwide tour, mostly balancing the demands of fame with his own code of ethics. He did his duty as he saw it. He was conscientious and polite and participated in causes he deemed worthwhile while avoiding anything that focused on him. When it was finally over they had visited 28 cities in 25 countries, and had been received by the Queen of England.

"I think Neil saw the results of being an idol when he researched Lindbergh's experience," said his friend Jim Lovell who commanded *Apollo 13*'s near disaster. "He didn't want his life to change. He decided to be very reclusive," Lovell explained. "That was just Neil's nature."

Jim Lovell, as did many of Neil's close friends, completely understood the boy from Smalltown, USA.

Others within NASA didn't.

The agency brass paid Neil to run a department in Washington and gave him the fancy title of Deputy Associate NASA Administrator for Aeronautics with an office and a view of the Capitol. As Neil said, "I went back to aeronautics from whence I came."

It was clear from the beginning Neil Armstrong wasn't pleased with the noise and fuss. "It was the NASA administrator asking if I would help him in that area, an area I felt comfortable with and had knowledge about," he added. "I was glad to have the experience, but I quickly came to the belief everybody should have to go to Washington and do penance.

"It was a frustrating place for me because so much coordination and greasing the skids goes on in Washington," Neil continued, "and I asked myself, 'What the hell am I doing here?'"

What NASA really wanted from Neil Armstrong was the persona of a used car salesman—a glad-hander—an agency star the agency could trot out, as needed. It was an assignment that solidified in Neil's mind the very living definition of divine punishment.

Neil couldn't stand it.

In 1971 he went back home and watched the moon-landing missions come to an end a year later. He was back among the farmlands of Ohio. He bought a dairy farm near the small town of Lebanon, and fulfilled his wish to teach. Neil took a post as an associate professor of aeronautics at nearby University of Cincinnati and worked hard as a teacher. He was comfortable being called professor, and for a decade he tried to do what was expected of him while avoiding becoming a public figure.

He found solitude in milking the cows, slopping the hogs, mending fences, and teaching kids what made it possible for machines to fly. Down at the Cape astronauts were eyeing the sky again. Nearly six years without a spacecraft following the July 1975 Apollo/Soyuz mission that flew the first international docking in Earth orbit, America was ready to return to space. Nothing like the machine the astronauts were about to fly had been seen before.

It was a winged spaceship the size of a jetliner and NASA was building four. They were named *Columbia*, *Challenger*, *Discovery*, and *Atlantis*, and each

Columbia: The first space shuttle launches. (NASA)

stood on its tail on its launchpad strapped to two towering solid-rocket boosters. A huge tank filled with more than a half-million gallons of super-cold fuels held them together. The boosters and the machine's three main rocket engines were to launch the winged craft into orbit like Mercury, Gemini, and Apollo. Using wings, rudder, and landing gear they were to touchdown on a runway much like a jetliner.

Called the space shuttle, it came with lots of promises when the first of them, *Columbia*, was rolled out. The media horde returned, settling on the same press site from which they covered Apollo. On April 12, 1981, the first space shuttle's countdown was under way.

On board ready to fly the new winged spaceship were a veteran astronaut and the commander of *Apollo 16* John Young, and a rookie astronaut Robert Crippen who was conspicuous driving a pickup truck made up principally of rust and worn-out parts. He was a friendly cuss—what women called handsome—a Texas roughneck most men wanted for a buddy and most women wanted for obviously something else. When the countdown neared its end, something never-before-seen happened.

Ignition began in a swift rippling fashion, a savage birth of fire as *Columbia*'s three main engines ignited, one after the other, creating a blizzard, a swirling ice storm shaken from the flanks of the shuttle's fifteen-story-tall super-cold external fuel tank. But the winged ship didn't move. It just sat there.

Just as the mighty Saturn V moon rockets had been held a few seconds to make sure all was running as it should, *Columbia*'s hold-down arms kept it firmly attached to its launchpad. Its three main liquid-fuel engines screamed and roared, and when the computers had sensed they were running well a rage of flame joined them from the ignition of the two giant solid boosters. *Columbia* leapt from its pad—the same launchpad from which Neil Armstrong, Mike Collins, and Buzz Aldrin had left for the moon. *Columbia* was climbing from its insanity of fire shattering the quiet of Florida's spacecoast.

Crowds were at the fences, on the causeways, standing on the beaches, and in the thickets and when two minutes had passed, *Columbia* kicked away its burnt-out boosters and sped like a homesick angel into Earth orbit. John Young

Space shuttle *Columbia* returns to land on California's high desert. (NASA)

and Robert Crippen, the gutsy fools, were grinning. They were in space where Crippen savored the joys of weightlessness while Young simply enjoyed being back. He needled Crip with one question, "Did you lock your pickup?"

Crip's disreputable pickup fell far short of anything worth locking and the two went happily about executing mission planners' short flight of 54 hours. The bosses wanted safety margins as wide as possible for the new machine and John and Crip checked and rechecked all its systems. When done the two fliers glided their new ship to a perfect touchdown on California's high Mojave Desert—the same runways where Neil Armstrong landed his X-15 rocket plane twenty years earlier.

Neil was enjoying life back on the farm but he could not help feeling a little twinge of missing it all as he followed the new space shuttles' flights. When the shuttle fleet approached its fifth anniversary, Neil was impressed the winged ships had flown 24 times.

"Spaceflight is becoming as safe as flying on a commercial jetliner," said some NASA executives and, with President Ronald Reagan's blessing, the agency went off and conducted a nationwide contest for a teacher to fly in space. They wanted a teacher to teach from orbit, and out of thousands who applied, they found the perfect candidate—a smiling, next-door girl clean of heart and spirit named Christa McAuliffe.

The teacher would ride aboard *Challenger*.

Some warned that NASA was overconfident, risking flight safety. The harbinger of that warning was rolling southward. The omen was a bitter cold wave freezing and crippling everything in its way.

America's girl next door, New Hampshire schoolteacher Christa McAuliffe. (NASA)

TWENTY-THREE

AN AMERICAN TRAGEDY

Florida's rare, bone-chilling freeze stiffened and split tropical flora as fire and smoke rose from smudge pots in citrus groves. It brought the nation's spaceport to a slow crawl in the predawn hours of January 28, 1986.

For the first time anyone could remember frost appeared on windshields. Icy fog formed above canals and lagoons. The living shivered. The disbelieving recorded 27 degrees before sunrise with not a single tropical insect moving. Birds accustomed to warm ocean breezes huddled. The space shuttle *Challenger* stood bathed in dazzling floodlights, seeming to ignore it all. Its metal and glass and exotic alloys unfeeling as the great ship of science rose 34 stories above its concrete and steel launchpad. Night slipped away and sunrise brought the first touch of warmth.

The space plane's crew of seven appeared on the launchpad. Among them was Sharon Christa McAuliffe with a smile as wide as her New England roots. She had been selected from thousands of applicants to be the "First Citizen in Space."

But McAuliffe wasn't going into orbit as a tested scientific or engineering member of the crew. The social-science teacher from space was going to teach Earthbound classrooms of awestruck students.

Following a morning with a stop-and-go countdown waiting for the temperature to rise, Launch Director Gene Thomas polled his team for a critical litany of last-second review. Every response was "Go!" Not a single call to halt the count as NASA commentator Hugh Harris reported the final moments. He spoke into a microphone that carried his report into officialdom and every media outlet worldwide. He watched the numbers shining brightly before him. Green and flashing, they gave him an update with each passing second, and as the seconds grew shorter he reported, "T-minus ten, nine, eight, we have main engines start, three two one . . ."

Ignition began as a coruscating fire, a sudden giant flash, and the towering space plane kicked free of its launchpad, spreading its rolling thunder and flames as Harris shouted, "Liftoff! We have a liftoff of the twenty-fifth space shuttle mission."

On board, the astronauts felt *Challenger* come alive and when the boosters ignited crew commander Dick Scobee shouted, "There they go, guys." Beneath him on the middeck Christa McAuliffe shouted words for her students into her tape recorder. She took just enough time to remind herself to grip her seat tightly for the ride that those who had gone before promised would be better than anything offered by Disney.

No one knew at the moment of solid rocket ignition, but something sinister was happening. Barely apparent beside the opening fiery blast, a puff of black smoke shot forth from the lower joint of the right booster. Almost as quickly as it happened, it was gone. Later examination revealed that the smoke had spewed from a sudden, tiny gap in a critical O-ring. Last night's freeze had robbed the critical seal of its ability to flex, to expand and seal. The puff of black smoke was *Challenger*'s death warrant.

High above and unaware they were in mortal danger, the astronauts shouted with excitement where the wind howled horizontally at hurricane speeds. *Challenger* pushed into Max-Q with determined power—but this flight was carrying a terrible flaw.

When the side-loads of the winds smacked into the right booster, they struck an already weakened rocket. The force of 84-miles-per-hour may have reinitiated or magnified the leak. Either way flames were now impinging on the external tank. The vehicle structure was compromised beyond its design limits.

There was nothing left to hold back the raging fire and enormous pressure that was generating the solid rocket's thrust. A spear of flame gouged through the small hole, carving an instant opening and spewing a blowtorch.

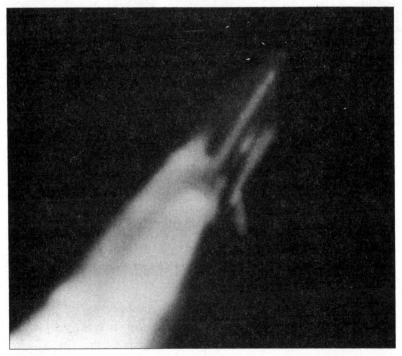

Challenger begins breaking up in its climb above its frozen launch site. (NASA)

Challenger was 58 seconds into its flight. Nothing could keep the winged space plane from coming apart.

Not one of the astronauts knew their right booster was already shredding itself.

The pilot's seat was on the right side of the spacecraft, nearest to its disintegrating booster rocket where suddenly a sheet of intense flame swept across pilot Mike Smith's window. In whatever instant of time was available to Mike, he knew something terrible was happening. He had just enough time to utter, "Uh-oh!" The cutting torch slashed through the lower half of the external fuel tank that stored liquid hydrogen. It collapsed and instantly disintegrated. Where there had been only blue sky pierced by bright flame below a climbing space shuttle, a hellish fireball grew.

Where there had been blue sky *Challenger* comes apart in fiery twisting smoke. (NASA)

Two corkscrew spears of white smoke spun twisting paths even higher, the rocket boosters flaming out of control. The instant fire in the sky continued to expand in a scattering of flaming debris, creating hundreds of burning, twisting fingers of smoke that seemed to be running from the growing conflagration.

In one ghastly moment, the very air over America's spaceport burned. Thunder echoed and boomed downward. It kept booming and thundering for the longest of unmeasured time. *Challenger* was breaking and shredding itself into millions of pieces, while beneath this sky of ominous groans, thin wailing cries and screams rolled upward from Earth to where *Challenger* died.

Neil Armstrong was devastated. He couldn't believe NASA had dropped the ball so badly, killing seven of its best. The very next morning the president was on the phone. Mr. Reagan was asking Neil to join the Presidential Commission on the Space Shuttle Challenger Accident. The president had asked former Secretary of State William P. Rogers to chair the commission, and now he was asking Neil to serve as vice chairman.

Neil Armstrong appointed vice chairman of the *Challenger* Investigation Board by President Reagan. (The White House and NASA)

The commission was given the job of learning in the next 120 days what happened to *Challenger*. Bill Rogers would work the politics, grease all the needed palms, and Neil would run the operational side—form committees and lasso investigators who knew where to look.

"I had enormous admiration for Bill Rogers," Neil would say. "He had a very good appreciation of Washington and what the needs of the public and press and the Congress were. He was to be 'Mr. Outside,' and he asked me to stay inside. I explained I didn't know the inner workings of the shuttle's systems, components, but I knew the program in a macro sense—objectives, techniques, and general strategies, and I knew the people who knew."

"That'll work," said Rogers, and Neil went out and grabbed the best hard-core investigators he could, not wishing to throw a wrench in NASA's own investigations. Neil knew in the long run both groups would be working together.

The locust arrived overnight. Hundreds of top names and known faces in the television world devouring everything in their way. This grunt in the field, who had covered every mission flown by American astronauts—56 at the time—suddenly had a whole bunch of best friends and nationally known talking heads to bow before.

In its wisdom NBC News decided my experience and well-placed sources could best be used for investigative reporting. I was off poking my nose in places where it came close to getting chopped off. I was eavesdropping on every meeting and skull session I could to learn what went wrong. Then I locked myself in my office and worked the phone, talking with other grunts, including those who turned the wrenches and cleaned up the messes, as well as supervisors and management types. I kept getting the same responses: no facts, only opinions.

A full day passed, and suddenly a brick hit me in the head: Sam Beddingfield, the man who had retired only a couple of weeks before as deputy director of space shuttle management; the same Sam Beddingfield who told Gus Grissom he didn't need a parachute because he wouldn't have time to put it on; the same Sam Beddingfield who Gus told "Put the parachute in my Mercury capsule anyway. It'll give me something to do until I hit." That Sam Beddingfield had all the experience and contacts needed to rub elbows with all the NASA brass on headquarters' fourth floor. I grabbed the phone.

"Hello."

"Sam, this is Jay Barbree."

"Yeah, Jay, what's up?"

"What'd you think's going on with management on the fourth floor?"

"They're running around, pointing fingers, protecting their asses," Sam offered.

"Most likely," I laughed, quickly adding, "Why don't you go down there and check it out?"

"I could," he smiled.

"You want a job?"

"Doing what?"

"You could be NBC News's space analyst for the *Challenger* accident."

"I could," he laughed. "It'd keep me outta the pool halls."

"It would at that," I agreed.

"Okay, I'll take a drive down to my old office—see what's going on."

"You're on the payroll, Sam," I told him, "So keep in touch."

"I will," he promised.

Sam Beddingfield sprung from the same roots my family did in eastern North Carolina. You could trust him with your children. Honesty was a way of life for Sam. He parked himself in the executive offices at NASA headquarters. He listened to everything so far learned. Most of the NASA managers simply thought Sam was still on the job. In the middle of the afternoon, January 30, 1986, two days after *Challenger* disintegrated nine miles above the Atlantic surf, Sam called me.

"I got it."

"You got what?" I questioned quickly.

"We lost *Challenger* because of a leak in a field-joint," Sam said flatly.

"An O-ring leak," I asked?

"That's it."

"For sure?"

"For sure."

"How do they know?"

"They have pictures," Sam said without hesitation.

"Pictures?"

"Pictures of the leak," Sam explained. "They show flame blowing out of the sucker like a blowtorch."

"Where did the pictures come from?"

"From a fixed engineering camera north of the pad."

"Away from our cameras? Where we couldn't see?"

"That's it."

"What did the leak do, burn into the tank?"

"Not at first," Sam explained. "It burned through a tank support structure before reaching the tank. Its fuel," he continued, "fed the flames—hell, they burned everything they touched."

"And just what caused the leak?"

"My best guess," Sam explained, "the O-ring was still frozen at launch. The sucker couldn't do its job."

"Makes sense," I agreed before asking again. "We can't see it on our launch tape?"

"No way."

"Can you get me those pictures, Sam?"

He laughed. "You trying to get me shot?"

"Not today," I told Sam, congratulating him again before hanging up and phoning a trusted source at the Marshall Space Flight Center in Huntsville, Alabama. She confirmed Sam's information and with two solid sources we were ready to break the story on Tom Brokaw's *NBC Nightly News*.

Our producer Danny Noa shoved a small model of the space shuttle in my hands and we opened the show that evening with the story.

Using the model I pointed to the suspected location. I answered Tom's questions, essentially reporting that the cause of the tragedy was the failure of O-ring seals in the aft field-joint of the right solid rocket motor. The extreme cold had rendered them inflexible. Frozen, the O-rings simply could not do their job.

No sooner had I left the air than the phone began ringing. Producers Geoff Tofield and Danny Noa were fielding most of the questions, but one came in I had to take.

"You looked pretty sharp for a farm boy tonight," Neil Armstrong said.

"Hey, Neil." I was grinning all the way through the line. "How's the milking going?"

"We're getting a couple of quarts," he laughed. "What can you tell me you didn't tell Brokaw?"

"Nothing more, really," I assured Neil. We talked about what I knew and I promised to keep him up to speed on what I learned. I did, and in less than four months Neil and the Presidential Commission on *Challenger* confirmed

Tom Brokaw and Jay Barbree breaking
what would later be voted the number
one story of 1986. (*NBC Nightly News*)

my report. In doing so the first man on the moon provided thorough leadership throughout the presidential commission's investigation, again setting another example to follow.

The commission found that the *Challenger* accident was caused by a failure in the O-rings sealing the aft field-joint on the right solid rocket booster. The failure was caused by pressurized hot gases and eventually flame "blowing by" the O-rings. Flame made contact with the adjacent external tank structural support, causing the structure to rupture the tank. The failure of the O-rings was attributed to a design flaw, as their performance could be too easily compromised by such factors as those low temperatures on the day of launch.

The report also determined the contributing causes of the accident were failure of both NASA and its contractor, Morton Thiokol, to respond adequately to the design flaw. The commission offered nine recommendations to improve safety in the space shuttle program.

Neil Armstrong and *Challenger* investigators inspect a space shuttle from beneath. (NASA)

Neil and his fellow commission members summarized their conclusions extremely well in the final report: "The decision to launch the *Challenger* was flawed. Those who made that decision were unaware of the recent history of problems concerning the O-rings and the joint and were unaware of the initial written recommendation of the contractor advising against the launch at temperatures below 53 degrees Fahrenheit and the continuing opposition of the engineers at Thiokol after the management reversed its position. They did not have a clear understanding of Rockwell's concern that it was not safe to launch because of ice on the pad. If the decision-makers had known all the facts, it is highly unlikely that they would have decided to launch *Challenger* on January 28, 1986."

The space shuttle's solid rockets were redesigned as the commission ordered and Neil returned to retired life. "I think our conclusions and findings were right on," he told me. "It was a very hard-working commission and our answers have never been effectively challenged. . . . It was a national tragedy," he continued, "but we learned a great deal from it, and the subsequent shuttle program benefitted. That was ours and spaceflight's reward."

September 29, 1988: Space shuttle *Discovery* sat on its launchpad. Five seasoned astronauts waited: Commander Rick Hauck, pilot Dick Covey, crew members Mike Lounge, Dave Hilmers, and George "Pinky" Nelson. They had been handpicked to return the rebuilt shuttle to flight after seven of their numbers had been lost.

Two hundred and fifty thousand other souls surrounded the spaceport to lend their support. Twenty-four-hundred members of the news media stood at the press site. They were there to report NASA's comeback from its worst disaster.

At 11:37 A.M., eastern time, *Discovery*'s main engines roared. Seconds later the twin solid rockets fired. The assembled thousands crossed fingers and gritted teeth as the two redesigned solid rockets lifted the five astronauts skyward, boosting the space plane and rocket combination straight and true. Two minutes later the huge viewing assemblage broke into wild cheers. The boosters, whose predecessors had been the primary cause of the *Challenger* accident, burned out and peeled harmlessly away from the shuttle and its human cargo. They were the first of 220 of the redesigned solid boosters that would be flown without the slightest problem until they and the space shuttle fleet were retired.

Possibly Neil Armstrong's greatest legacy following his life of flight as a pilot, engineer, and investigator was the way in which he handled his extraordinary life, which resulted in many of those who would drive aircraft and spacecraft to regard him as one of the best of all times. Neil never thought of himself as special, but everyone else did.

President Ronald Reagan opened an awards ceremony in the White House Rose Garden with the announcement, "America is back in space."

TWENTY-FOUR

SPACE SHUTTLE AND BEYOND

NASA entered the final decade of the twentieth century fully recovered from its worst spaceflight accident with the agency building on the pioneering work of Neil Armstrong and the astronauts of the space industry's first three decades. A couple of pretty fair space shuttle drivers Robert "Hoot" Gibson and Charlie Precourt both walked in the shoes of Deke Slayton as chief astronaut. Both played major roles in one of Neil's oldest dreams—building a permanent space station.

The International Space Station would in some minds be the beginning of an orbiting space city, a gravity-free outpost where Earthlings could multiply, raise families, live longer, and produce the stuff and foods needed for self-sufficiency in space.

But as Gibson, Precourt, and a handful of others knew, the station's primary job would be to teach humankind how to survive in space. NASA already had a small taste of operating its own space station in 1973 when it used rockets and spacecraft modules leftover from canceled Apollo flights. The agency had launched three separate crews of three astronauts each to spend up to nine months aboard a station named Skylab. The astronauts proved humans could work in space. But soon after the Russians lost the moon race they, too, got serious about space stations. They launched a series of Salyut laboratories, each carrying two or three cosmonaut crews who could stay in space for months.

There was that brief thaw in the Cold War in 1975 when Deke Slayton's heart was deemed strong enough for space flight. Tom Stafford commanded and Vance Brand rounded out the three-man crew of Apollo-Soyuz. They drove the last Apollo to a rendezvous and docking with the Russians' Soyuz. Eleven years later, in 1986, Russia sent into orbit what was, to many who made space stuff work, the first real space station. It was called Mir, and cosmonauts stayed aboard their home in the sky for up to a year.

In 1991, as the Cold War ended, Presidents George Bush and Mikhail Gorbachev signed an agreement calling for the first U.S.-Soviet manned space flights since Apollo-Soyuz. It was to include a visit by a U.S. astronaut aboard a Soyuz-TM spacecraft, and a flight by a Russian cosmonaut on a shuttle mission. A major milestone of this agreement occurred on June 29, 1995, when Hoot Gibson and Charlie Precourt docked space shuttle *Atlantis* for the first time with Mir. They exchanged crews and checked out the Earth-orbiting community.

Astronaut Charlie Precourt was one of those directly responsible for helping bring Russia on board as a full member in the U.S.'s plans for building the International Space Station. He reflected on how the first docking with Mir was a beneficiary of the pioneering work of Armstrong and the Apollo astronauts:

> Docking the shuttle with the existing Mir station would be our first step in learning to work together to build the International Station. We had done lots of rendezvous missions with the shuttle to retrieve satellites— like when we repaired the Hubble Space Telescope—but we hadn't yet physically docked with another spacecraft. Mir and Shuttle each weighed over 200,000 pounds and we knew lots could go wrong if we didn't prepare correctly. Neil's experience was a critical contributor to our success. We studied the results of the Gemini and the later Apollo dockings and we had procedures for every conceivable combination of thruster issues as well as a number of other potential failure modes. Neil's first-ever docking of Gemini with the Agena had to be aborted because of a stuck thruster. His quick reaction saved the mission and likely his life. That experience helped every subsequent mission of its kind. Hoot and I, along with our flight engineer Greg Harbaugh, put hundreds of hours into simulations just as Neil did, to test and adjust our plans.

Precourt paused and then continued: "When the day came it went off beautifully. Hoot made contact with Mir at 1.2-inches-per-second in perfect alignment with the Mir docking mechanism. We were within a second of the planned docking time."

The man who was counted among the Air Force's best test pilots said honestly, "Our success was largely due to Neil having paved the way," and Precourt's praise for Neil Armstrong didn't stop there. The chief astronaut went on to say:

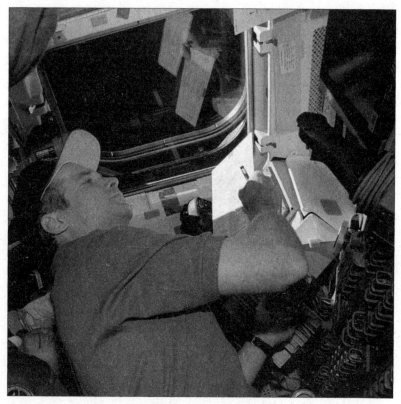

Charlie Precourt commanded two dockings with Mir and is seen here driving the shuttle to a perfect arrival with the Russian space station. (NASA)

In the astronaut office of the shuttle era, we all admired Neil's accomplishments as a pilot. We had all been inspired to pursue careers in aerospace and strive to become astronauts watching the first moon landing in 1969. Neil's abilities as a pilot were what we all aspired to.

But what was most amazing to us was how well he performed as a pilot on so many "firsts." We all knew it was challenging enough to fly a spacecraft even on flights downstream of the first, but Neil made all his firsts look like he'd been there many times. And yet as inspiring as Neil was to us as pilots, perhaps most inspiring to those of us who flew as shuttle astronauts was Neil's character and humility. He never wanted to talk about Neil. He would always vector the conversation to the team, and the lessons of the mission, or the machines. His eyes would

The space shuttle and Mir securely docked. (NASA)

light up whenever the conversation was about comparing experiences with different aircraft. Neil was the ultimate example of a gentleman, and we all looked up to him for his strength of character.

Other astronauts also like to tell stories about Neil, like the time retired astronaut Franklin Chang-Diaz had been asked to give a talk on his work on a plasma propulsion system during one of the astronaut reunions. Plasma propulsion many feel is the gateway to interplanetary travel. Chang-Diaz remembers, "After the briefing, while I was packing up my computer, one of the attendees came to me and said, 'Hi, I'm Neil Armstrong. I really enjoyed your presentation and I am glad you are working on this—it is what we need to get to Mars.' My jaw dropped. I actually had never had the opportunity to meet one of my all-time heroes. I found myself at loss for words."

Similarly, shuttle commander Brian Duffy relates being with Neil at the 30-year reunion of the *Gemini 8* team in Houston, an experience that made a real impression: "I had heard that Neil was a fairly private person who avoided the limelight that accompanied his feats," Duffy said:

I was not disappointed to learn firsthand that he was exactly that, but very personable as well. The reunion brought people who had worked together as a great team, and let them reconnect after many years. Neil was wearing a lightweight summer suit because of the heat, and his eyeglasses had large lenses that tinted automatically in the bright daylight. He made some brief remarks to the mostly older assembled group, and they ate up every word he said. He was obviously very respected by all.

After a few hours, it was time for him to head to the airport for his evening flight home. We proceeded to the counter for Neil to check in. The young lady behind the counter, perhaps in her early twenties, took the reservation Neil handed her and proceeded to print out a boarding pass. From the side, I was watching for any sign that the young woman recognized perhaps the most well-known name in the world as the gentleman with the tinted glasses and old briefcase that stood in front of her. There was nothing. No recognition. She calmly folded the boarding pass and documents and handed them back over the counter to Neil and said, "Have a nice flight Mr. Armstrong." Neil softly said, "Thank you," and picked up his things. Respecting Neil's humility, I never said anything to the agent; not that I didn't want to let her know whom she'd just met!

Charlie Precourt added:

For those of us fortunate enough to have flown the space shuttle, Neil Armstrong was our inspiration, our role model, and the pride of our astronaut corps. It would be difficult to imagine a more appropriate representative of humankind to be our first to set foot on the moon. He accomplished the mission with the highest level of skill and has since represented all of those who contributed to that feat with the greatest of humility and dignity.

One of my most rewarding assignments was when I was privileged to select, train, and work with many shuttle and International Space Station crews. All had their own particular challenges to prepare for missions. It wouldn't be a stretch to say that whenever we would question ourselves about how to handle a situation, in flying or in life, we had Neil's example to lean on.

Space shuttle trips into space were running like a well-oiled clock when Neil began needling his jungle training buddy John Glenn. Neil knew the old space icon was getting restless, knew he had wanted to go back into space since Kennedy grounded him for being too much of a hero. "Now's the time for you to go, John," Neil said. "You're a big bad United States Senator. Who's going to stop you?" Neil teased.

John finally decided Neil was right. He'd had enough of Washington politics, and had retired from the United States Senate after serving 24 years. But there was one small problem! John was 77 years old. Many asked, "Can a septuagenarian handle it?"

"That's precisely the point," argued John. "I should go up and let the scientists take a look at this aged body, see if seniors can function better in space than here in the gravity of Earth."

Chief astronaut Charlie Precourt smiled and gave his okay and NASA Administrator Dan Goldin gave John the green light. Space shuttle *Discovery*'s Commander Curt Brown and its pilot Steve Lindsey welcomed the shining knight of earlier years with open arms, and the spaceport swelled with grey-haired pride.

On October 29, 1998, anchor Brian Williams and this old broadcaster were before the NBC cameras reporting John Glenn's launch into a blue and happy sky. Neil and I and those who admired and loved John Glenn were never more proud. We prayed and wished him luck and spent nine days watching this 77-year-old never miss a step.

In orbit John proved to be the champ we all knew he was by taking care of his assignments, having fun, and doing a little rocking and singing. Commander Curt Brown reported, "Let the record show John has a smile on his face and it goes from ear to ear."

Glenn took the microphone and added, "Hello, Houston. This is PS 2. They sprung me out of the mid-deck for a little while. This is beautiful. The best part—zero G—and I feel fine," same as he said from *Friendship 7* on America's first orbit 36 years earlier.

Medical research during Glenn's mission included a battery of tests on John to research how the absence of gravity affects balance and perception. The research also took a long look at the immune system response, bone and muscle density, metabolism, blood flow, and sleep.

John Glenn's second flight at age 77 did much for extending the retirement

John Glenn, 77, training for his space shuttle flight. (NASA)

age. On November 7, 1998, at 12:04 P.M., eastern time, John Glenn and crew aboard *Discovery* touched down on its Florida landing strip.

Less than an hour later, with all housekeeping chores completed, John Glenn strolled off the shuttle *Discovery* seemingly without a care in the world. The lack of gravity in space weakens astronauts' arms and legs, and it takes some, even those in their thirties, days to return to normal. It was believed that Glenn would need a wheelchair.

Well, forget about it! John walked by me and winked, and I hit a smart salute and hid a couple of tears. He'd just brought us aged ones hope we could keep on marching through life with purpose.

Neil had no better friend than John Glenn and when he saw John doing his

An hour after landing from his nine-day spaceflight, Chief Astronaut Charlie Precourt and NASA Administrator Dan Goldin match John Glenn stride for stride for his crew walkaround inspection. (NASA)

"nine-day spaceflight walkaround inspection" only an hour after landing, Neil didn't stop smiling for days.

But in spite of the joy John Glenn's second flight brought him, the last decade of the twentieth century wasn't without its unhappiness. Neil spent most of it with family when all wasn't well between him and Janet. They'd grown apart, and on April 2, 1994, his wife of nearly four decades divorced him. For five years he was lost until he met and married Carol Knight, a widow. They set up housekeeping in the Cincinnati suburb of Indian Hill.

About the only thing Neil ever said to me about Carol was he knew he wasn't in her league. She has a smile as wide as Ohio and warmth as big as Texas. No person ever met Carol that didn't like and appreciate this beautiful addition to our species.

Carol and Neil were introduced and taken to dinner by mutual friends. Neil being Neil took his own sweet time getting back to Carol. Hell, he'd decided to propose to Janet two years before he asked. When it came to Carol, he kept asking himself why this magnificent woman would be interested in an old worn-out pilot like me.

He could not come up with a plausible answer, but he couldn't forget her, either. He finally phoned. Quietly he said, "Hello, this is Neil."

"Neil, who," she answered in a rush.

Embarrassed, Neil reminded her who he was. She immediately told him she and her son were having trouble cutting a tree down in their backyard and she didn't have time to talk. Neil, realizing he could be a hero to the rescue, spoke up excitedly saying, "I can take care of that."

Carol rushed her good-bye and ran back to the job at hand.

Only minutes had passed when there was a ring at Carol's door. Her son answered, only to run back to his mother, shouting, "Mom, you're not going to believe who's at the front door with a chain saw."

Neil immediately sensed the teen's reaction had something to do with *The Texas Chainsaw Massacre.* Carol had not told her son she and Neil had been to dinner with friends and that he was coming over to help with the tree.

Carol and Neil would have 14 years together during which the century turned, and Neil was glad to see construction was moving ahead on the International Space Station.

The orbiting outpost was to be as large as two football fields set side by side. But on February 1, 2003, construction stopped. The space shuttle *Columbia* and its crew of seven were lost, again to an aura of what might best be called arrogant complacency.

The flight by the oldest member of the shuttle fleet had nothing to do with the space station. It was in its own orbit for 16 days where it could have only been seen by national security assets. Had it docked with the space station, or had those in charge of its flight decided to look, the hole in its right wing surely would have been detected. The hole got there during its launch when insulation cascaded from the *Columbia's* external fuel tank and ripped the hole in its right wing.

While in orbit, an overconfident NASA failed to inspect the ship following questions raised from viewing launch video. When the oldest shuttle and its seven returned to Earth, reentry heat devoured the space plane and breached a wing. *Columbia* and its astronauts disintegrated over Texas and Louisiana while penetrating Earth's atmosphere.

Neil was again devastated.

The Columbia Accident Investigation Board recommended the space shuttle fleet be grounded following completion of the International Space Station. Neil and a disheartened space family watched the space shuttles become museum pieces.

NASA was virtually dismantled because of a lack of interest and indecisive leadership. Neil joined the last man on the moon Gene Cernan and *Apollo 13* Commander Jim Lovell in writing an op-ed on the drifting space agency. The three Apollo astronauts saw fit to give their op-ed to me; Brian Williams broke the editorial on his number one news program, *NBC Nightly News*.

Neil, Jim Lovell, and Gene Cernan traveled the planet preaching what we must do if Earth and its miracle of life are to survive.

During this period, as Buzz Aldrin neared his eightieth birthday, Neil's partner on the moon was approached by a six-foot-something, obnoxious conspiracy theorist. This clown didn't believe Aldrin had walked on the moon. The jerk got right into Buzz's face with a microphone and camera. He shouted, "You are the one who said you walked on the moon and you didn't. Talk about the kettle calling the pot black."

"Will you get away from me?" Buzz asked not so politely.

"You are a coward, a liar, and a thief," the harasser shouted.

Having to reach up, Buzz threw a swift punch startling and sending the guy backward. The idiot called to his cameraman, "Did you get that? Did you see what he did? Did you get that?"

Those watching laughed as the conspiracy theorist ran, clearly identifying who the coward really was.

Soon Buzz's astronaut buddies like Wally Schirra began calling Aldrin "Rocky." When that same idiot approached Neil asking him to swear on a Bible he was holding that he walked on the moon, Neil smiled pleasantly. "I would, but I'm sure like you your Bible is a fake."

A couple of working scientists on Discovery Television's *MythBusters* took the conspiracy theorist's so-called evidence, and with scientific tests debunked the theorists' claims.

In July 2009, NASA's first lunar scout for the twenty-first century, the Lunar Reconnaissance Orbiter, took images of the Eagle's lunar module descent stage where Neil and Buzz left it on the Sea of Tranquility.

Apollo 14's landing area showed the Lunar Reconnaissance Orbiter a faint trail of Alan Shepard's and Edgar Mitchell's two-mile round-trip walk to Cone Crater. You can see clearly where they pulled their "rickshaw" tool and collection carrier. *Apollo 17*'s picture is so detailed experts can tell which way the wheels were pointing on Gene Cernan and Harrison Schmitt's lunar rover.

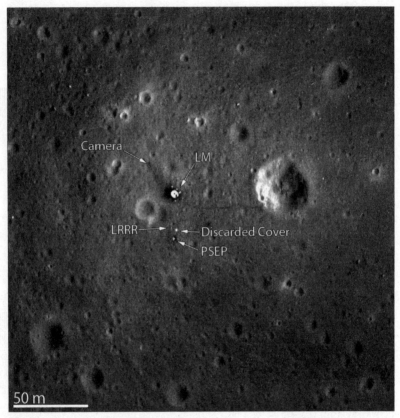

Apollo 11: Eagle and experiments still sit where Neil and Buzz left them. (NASA)

Following these pictures 40 years later, the conspiracy theorists were left with one claim: The United States Congress and President Barack Obama approved $200 billion for NASA to build and fly the Lunar Reconnaissance Orbiter to fake these pictures.

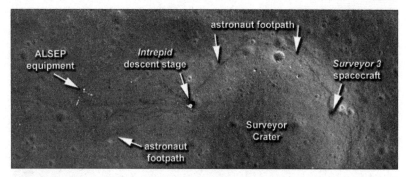

Apollo 12's lunar module descent stage along with Conrad's and Bean's footprints are still there. (NASA)

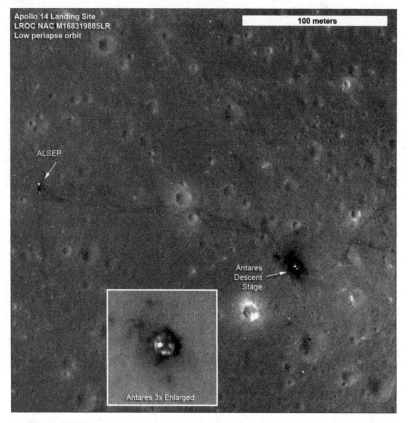

Apollo 14 Landing Site
LROC NAC M168319885LR
Low periapse orbit

100 meters

ALSEP

Antares
Descent
Stage

Antares 3x Enlarged

Apollo 14: Rickshaw tracks, lander, and experiments left by Alan Shepard and Edgar Mitchell are clearly seen. (NASA)

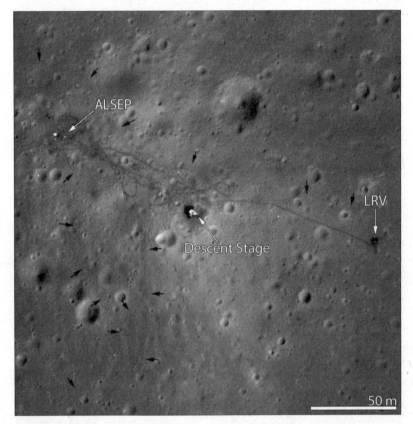

Apollo 15: First lunar rover tracks are clear as is the lunar rover where Dave Scott and Jim Irwin left it. (NASA)

This reporter covered every mission flown by American astronauts for NBC News. We broke the cause of the *Apollo 1* launchpad fire that killed three as well as the cause of the *Challenger* accident that killed seven. If NASA had convinced 400,000 Apollo workers to lie and if the agency had slipped their fraud by the Russians, the British, and the Chinese who were tracking America's nine trips to the moon, I promise you the agency would have never gotten a lie by me nine times. Why wouldn't NASA stop after succeeding once? Why would they risk getting caught lying another eight times?

More important: Did NASA lie about the *Apollo 1* launchpad fire? Did the United States of America kill three of its astronauts just so it could create a

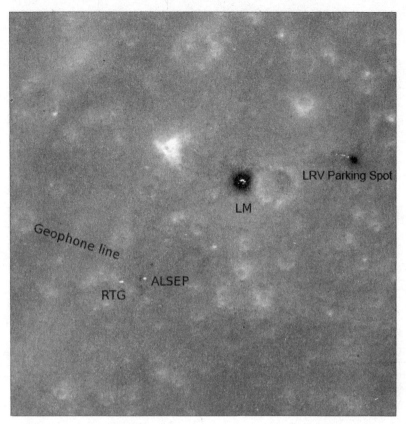

Apollo 16: Lunar rover is parked where it was left by John Young and Charlie Duke, to the right of the lunar lander. (NASA)

falsehood about placing astronauts on the moon? Only an idiot could believe that.

Sad but true: Neil Armstrong spent his last days disappointed that America's space program had been abandoned.

He contributed wholeheartedly to five stories we wrote for NBCNews.com entitled "Space in the 20Teens." The last of the features ran only a month before Neil left us. He died August 25, 2012, following complications from heart by-pass surgery.

Apollo 17: We see it all. The parked moon buggy, the experiments, the lander, and the tracks left by Harrison Schmitt and the mission's commander Gene Cernan, the last astronaut to walk the lunar landscape. (NASA)

No greater man walked among us. No better man left us informed answers. Neil taught us how to take care of our Earth-Moon system.

His thoughts are next.

The Hubble Space Telescope in orbit, cataloging the universe. (NASA)

TWENTY-FIVE

———

THEN, NOW, AND TOMORROW

First, there was nothing.

Then, 13.82 billion years ago the universe simply burst into existence, a "singularity"—a huge release of energy that would be called the Big Bang. It happened in total darkness because light didn't yet exist—even stranger, neither did space.

Today, the Big Bang is accepted. Most astronomers and scientists believe it was just a very tiny ultra-hot particle of energy before exploding in the greatest of all explosions; it expanded, inflated into existence, getting bigger and cooler with each passing moment. It created matter in the form of countless trillions of subatomic particles—the first stuff there ever was. Then it cooled! The fog cleared around 400,000 years following the Big Bang and the universe became visible. It was no longer a cloud of gas. Galaxies and quasars began forming in single points of light—uncountable billions of them spreading a vast web in every direction.

That evolution of our universe continues today.

Hubble's deep-field distant galaxies. (Hubble and NASA)

The first suggestion that the unimaginable explosion occurred was by a decorated World War I artillery officer, ordained a Jesuit Belgian priest in 1923.

His name was Georges Lemaître. He met with the pope of his time, Pius XII, and convinced him not to talk about creation any more. Lemaître felt strongly that one should not mix science and religion. He told the pope, "As far as I can see such a theory remains entirely outside any metaphysical or religious question. It leaves the materialist free to deny any transcendental Being. For the believer, it removes any attempt at familiarity with God. It is consonant with Isaiah speaking of the hidden God, hidden even in the beginning of the universe."

The pope agreed.

Lemaître referred to the "singularity" as a "cosmic egg" exploding at the moment of creation. He published in 1927 what scientists now call the Hubble Constant, which tells us how fast the universe is expanding. Later, Sir Fred Hoyle, an English astronomer known for his contribution to the theory of stellar nucleosynthesis, coined the Big Bang, originally a pejorative term. Hoyle believed in a static universe. Lemaître received a Ph.D. from MIT, and met with Einstein a number of times. Einstein originally agreed with the mathematics but not Lemaître's physics. Later the formulator of the theories of relativity called his rejection of the idea of an expanding universe, "the greatest mistake I ever made." At a meeting in Princeton in 1935, Einstein generated standing applause for Lemaître, referring to his work as "the most beautiful and satisfactory explanation of creation to which I have ever listened."

The universe's age of 13.82 billion years was determined in recent years by satellite measurements of the cosmic microwave background. These measurements also detected thermal residue of the Big Bang, making Lemaître's theory even more credible.

Neil Armstrong, as did most, questioned what prompted the singularity; where did the smallest of energy particles come from? Not that Professor Armstrong necessarily disagreed with accepted theory. He had questions. He just wanted to know if the singularity was simply an accident, or an act of some great force of nature, or if a supreme being orchestrated that big bang.

As of this writing no one will or can answer Professor Armstrong's basic question.

In his never-ending research Neil found that nine billion years after the "singularity," when galaxies were being born and billions of stars were burning themselves into dead matter, an ancient star had exploded. It littered our Milky

Way Galaxy about 26,000 light years from its center with whirling clouds of the materials it had made while it lived.

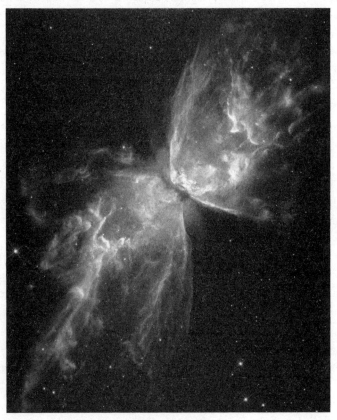

This photo of Nebula NGC 6302, the Butterfly, is an example of how our own solar system was born. (Hubble and NASA)

These whirling clouds of dust are called nebulae. They are very beautiful. But this one was different. It was the nebulae that gave birth to our solar system. It was nitrogen and oxygen and iron, and then the tireless forces of gravity pulled it all back together—the heavy engineering that produces planets had begun.

Vast spirals of dust gathered. At the center of one of these spirals a rocky planet that would be called Earth was taking shape. It was built from stardust and assembled by gravity. Within 100 million years it had grown into a giant ball sweeping up billions of tons of celestial materials. And at the heart of our nebula the pressure and temperature of this ball of hydrogen gas had become so great that the atoms were beginning to fuse. A new star, our sun, was coming to life.

As our sun ignited it gave off a huge blast of solar wind—a gust of energy that blew all the remaining dust and gas left over from the nebula out to the edge of our solar system.

In the outer reaches the huge gas planets Jupiter, Saturn, Uranus, and Neptune took up their orbits. Farther in—denser, rocky planets were fighting to survive.

There were about twenty and when there was nothing more to sustain them, things got nasty. They orbited the sun, their gravity affected each other, and they began to collide. With each collision they devoured one another, and over time the twenty became four: Mercury, Venus, Earth, and Mars.

Neil's research suggested Earth consumed about ten of its neighbors. The results were an incredibly hot planet—about 8,500 degrees Fahrenheit. Had there been any living thing on Earth then it would have disappeared in an instant—just a brief puff of steam and ash.

So it appeared all life had to do was wait for Earth to cool. Then 50 to 150 million years after its birth it seemed our young world had things under control. But it didn't. From deep within the solar system a large planetoid was headed Earth's way. It was on a collision course and it was unbelievably massive— possibly the size of Mars itself.

Had there been humans on Earth then they would have spotted this intruder millions of miles out. They would have watched it grow in size night after night. Soon it would have filled the sky. There would have been no escape, no reprieve. Instantly they would have been staring doom in its face as—at an oblique angle—the planetoid squashed the young Earth.

Our virgin planet reeled violently from that catastrophic blow. The planetoid exploded into the hottest of hot debris as it gouged a terrible, great wound in Earth—its mighty fires speeding on, leaving behind a flaming, quake-wracked planet. The mixed remnants of that titanic blast whirled back into space, there to be grabbed by Earth's gravity.

Most of the heaviest elements from the planetoid, especially its iron, remained with the now-molten Earth, beginning a long settling motion to the core of our world-to-be. As Earth cooled, it became a planet much different from the one it had been before the collision. Like Mercury and Venus without their moons, Earth had rotated sluggishly. But that terrible impact sped up our planet to one full rotation every 24 hours.

Again a harvest had been reaped. The flaming, vaporized planetoid hurled away from Earth, now settled into orbit. In the billions of years to follow it coalesced under its own gravitational attraction into the moon we see today.

Modern humans appeared on Earth about 200,000 years ago. Soon our ancestors were out of their deep African Eden, moving about, exploring, wondering where the moon came from. Then in the middle of the seventeenth century, Galileo and other astronomers fashioned crude telescopes and clearly saw on the lunar surface bright highlands and darker plains with endless overlapping craters.

Earth's great universities joined the telescopic studies in the eighteenth and nineteenth centuries. Then, like neighbors everywhere, a lunar visit in the twentieth century seemed the neighborly thing to do.

Neil Armstrong and Buzz Aldrin were first to land and walk on the moon. Within a period of four years, 24 Americans had visited our nearest neighbor. Some of them flew twice. Half of those 24 rode their landers down to the lunar landscape where they walked and drove on that small world.

Had Russia sustained its early lead the number of humans going to the moon might have increased greatly.

The Apollo trips alone leapedfrogged humans 50 years ahead in science and knowledge. But what most astounded Neil was not that we went to the moon, but that we didn't stay.

At this writing it appears at least one half-century will have passed before humans leave Earth orbit again. In Neil Armstrong's words, "I find that mystifying. It's as if sixteenth-century monarchs proclaimed that 'we need not to go to the New World again, we have already been there.'"

Neil spent his last years convincing our species of the importance of exploring.

He was guided by one absolute.

From the moment he read the words written in the nineteenth century by the Russian visionary Konstantin E. Tsiolkovsky, Neil had a clear understanding of where humankind should be going.

Tsiolkovsky is credited as being the first to envision the concept of using rockets for space travel. In a simple elegant use of language, the Russian scientist/schoolteacher saw the future, saw what humans must do and where they must go if they were to survive.

"Earth is the cradle of the mind," wrote this self-taught man reaching for tomorrow, "but one cannot live in the cradle forever."

Tsiolkovsky's words became part of Neil and he never felt their meaning more strongly than when he first left the cradle and enjoyed the freedom of sustained weightlessness aboard *Gemini 8*. Neil had marveled at those moments moving over oceans and seas with their surface glassy and undisturbed. He happily recalled seeing a shining streak, a single wirelike gleam of sunlight reflecting off a long, straight railroad track running through green fields and vast tracts of farmland for miles. Awed, he saw cities that were dark patches, huge rivers that were, from Neil's vantage, sparkling ribbons winding through the countryside.

But it was the night that held him most in wonder. Ironically, the darkness brought with it what was invisible during the day.

Cities sparkling and gleaming with multicolored lights, a swarming of neon illuminations, brilliantly lit streets, buildings ablaze from neat rows of glowing windows, man-made oases of color and brightness connected by long tendrils of highways marked with inching headlights.

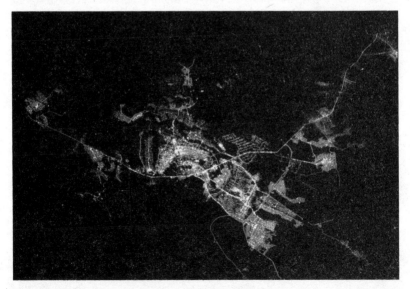

City streets and lit buildings as seen from Earth orbit. (NASA)

Aurora borealis as seen from Earth orbit. (NASA)

Then, silently, magically, glowing colors would rush down from arctic regions, the aurora borealis—electrical charges ignited by the sun in the upper atmosphere of Earth.

From *Apollo 11*, from near the end of his journey across the Earth-Moon system when they entered full darkness, Neil had stared, unblinking, at the illuminated objects and the companion Magellanic Clouds in his own Milky Way—a Milky Way full of dust and gas with its halo of dark matter. More than 90 percent of its mass contained an uncountable array of suns and nebulae and supernovae and whirls of stars within whirls of lights and colors, all members of our great pinwheeled galaxy of which we of Earth are merely one tiny member.

Neil Armstrong was a man humbled, awed, and grateful for what he had been privileged to see. He understood he was seeing clearly into tomorrow, focusing on life and time itself. Understanding that life was indifferent, realizing that time is a dimension measured only within the mind.

He was most aware that Earth was the human-species spaceship and that it would one day pass into history. No one knew when, but all knew it was inevitable.

That end should be eons away. But if another asteroid or another incurable disease decides to visit our planet, humans could be wiped out within a few

Neil Armstrong saw the Magellanic Cloud from *Apollo 11*, as shown here in this picture by Mount Hood Community College's Planetarium. (Mount Hood Community College Planetarium)

years. But if our leaders should have the foresight to populate the moon and Mars, even beyond, then life would go on.

As Neil had come to face, that wasn't likely.

What is and has been missing, and promises to continue at the rudder of the America Neil knew, is our lack of vision for the future. Leaders in both houses of Congress, but mostly Presidents Carter, Reagan, Bush, Clinton, Bush, and Obama, have kept astronauts locked in Earth orbit. These chief executives only gave America's space program short shrift. From 1981 to 2011 135 space shuttle missions were launched while America's leaders stood on the sidelines with little comment. The space shuttle flights were really only astronauts exploring the first onionskin beyond the surface of their own planet.

When President Ronald Reagan's daughter Maureen visited Cape Canaveral she only wanted to know, "When are you sending Sally Ride to the moon?" She wasn't at all happy to learn NASA couldn't send astronauts higher than 300 miles with the space shuttle. Today she would be sorely pissed to learn NASA can't send a live flea to the moon and bring the speck of life back still breathing. The agency doesn't expect to have hardware that can fly a living being beyond Earth orbit before 2021 or 2022.

With an international crew of six, the magnificent International Space Station orbits Earth. (NASA)

Wow, you say! But the truth is NASA hasn't been capable of sending astronauts beyond Earth orbit since 1972—an expected gap of a half-century. That's criminal! Yet the agency could explore farther out if our dysfunctional federal government would take a few taxpayer dollars from a couple of meaningless pet projects. But it won't. So for now to tend the International Space Station, NASA must keep buying its astronauts seats on Russian spacecraft. We will endure that humiliation until a commercial company or two can human rate their spacecraft for low Earth orbit taxis to carry astronaut crews to the station. That should be by 2017 or 2018, but none of these planned commercial spacecraft will be equipped to scratch a single particle in deep space. It'll just be the same-old-same-oh.

What Neil and many other visionaries long recognized is that if humankind is to continue its destiny as explorers we must conquer the problems of deep space travel in measured, incremental distances by first settling and fully understanding our own neighborhood—the Earth-Moon system.

Neil was aware, as most others were, that we could send spacecraft to Mars without life on board. We have done so with great unmanned NASA programs like its car-size rovers Curiosity and Spirit. They moved over the Martian landscape sending back information and pictures while two spacecraft did the same circling the red planet.

But Neil also knew that until science developed the means to travel at faster interplanetary speeds, Mars-bound astronauts would arrive as blubbering idiots from radiation poisoning—no ship headed to Mars today could carry enough lead to protect its crew.

What's needed is a lightweight radiation protection device like a small magnetic field. Neil was convinced that to develop such equipment we needed first to explore and settle our Earth-Moon system. This would teach us how to move on to more distance places like the Lagrangian points—those points in space affected equally by the gravity of two larger objects like the Earth and the moon so a smaller object can be left in one place.

But until we learned how to live in and master our own unaided personal survival while dealing with the hazards of space, Neil said, "We need the 3s. We need to stay within 3 seconds of communicating with Mission Control for its invaluable aid, and we need to stay within 3 days of returning to Earth.

"For deep space journeys we need to really know what we are doing," Neil explained. "We need to be masters of our own survival."

He went on to point out that Earth and the moon make up one celestial system. Neither world could survive without the other. It is the center of this dual system, rather than the center of Earth itself, that describes an elliptical orbit around the Sun in accordance with Kepler's laws. It is also more accurate to say the Earth and moon together revolve about their common center of mass than to say the moon revolves about Earth. This common center lies beneath Earth's surface, about 3,000 miles from our planet's central point.

"If the Earth-Moon system is our home, shouldn't we be caretakers of both worlds?" Neil asked. "You wouldn't manicure your front yard and leave your back to become overgrown with weeds. Shouldn't exploration of our own place and the Earth-Moon system's Lagrange points (all five of them) be our baby steps? Wouldn't it be silly to go trotting off to asteroids and Mars, millions of miles away, before we knew our way around our own neighborhood—before we really solve the problem of space radiation?"

Neil believed during his final years that NASA was going nowhere fast. He worried that the space agency no longer had a rocket of its own, leaving no direct way for astronauts to go from their own launchpads to the space station. He feared the international outpost could experience a major failure with little support from the country that assembled it in orbit.

In congressional testimony Neil testified:

> Americans have visited and examined six locations on Luna, varying in size from a suburban lot to a small township. That leaves more than 14 million square miles yet to explore.
>
> The lunar vicinity is an exceptional location to learn about traveling to more distant places. Largely removed from Earth gravity, and Earth's magnetosphere, it provides many of the challenges of flying far from Earth. But communication delays with Earth are less than two seconds, permitting Mission Control on Earth to play an important and timely role in flight operations.
>
> In the case of a severe emergency, such as Jim Lovell's *Apollo 13*, Earth is only three days travel time. Learning how to fly to, and remain at, Earth-Moon Lagrangian points would be a superb precursor to flying to, and remaining at, the much farther distant Earth-Sun Lagrangian points.
>
> Flying to farther-away destinations from lunar orbit or lunar Lagrangian points could have substantial advantages in flight time and/or propellant requirements as compared with departures from Earth orbit.
>
> Flying in the lunar vicinity would typically provide lower radiation exposure than those expected in interplanetary flight. The long communication delays to destinations beyond the moon mandate new techniques and procedure for spacecraft operations. Mission Control cannot provide a Mars crew their normal helpful advice (while landing) if the time delay of the radar, communications, and telemetry back to Earth are minutes away.
>
> Flight experience at lunar distances can provide valuable insights into practical solutions for handling such challenges. I am persuaded that a return to the moon would be the most productive path to expanding the human presence in the solar system.

When President George W. Bush decided to bring the space shuttle program to an end on the advice of the Columbia Accident Investigation Board, he put in its place Project Constellation.

Constellation would have given NASA two sets of rockets with emergency-abort systems and safety measures never in place on the space shuttles.

An early version lifted off in October 2009. It worked as advertised. But despite that successful flight and the fact that the new rocket offered the shortest and least costly route to keep American astronauts flying, Constellation was canceled.

Three years of infighting and knock-down-drag-out brawls failed to produce a replacement. Then, those who tossed Constellation away in the first place, hurried back to the waste dump and picked up its pieces. At the direction of Congress, NASA reinstated the big heavy-lift Constellation rocket—the first rocket since Apollo that could boost astronauts beyond Earth orbit. It is now the heavy-lift SLS (Space Launch System) with the development of an Apollo-type spacecraft called Orion.

When I learned NASA was adopting Neil's thinking—the agency was drawing up plans to reach deep space in increments—I told Neil. I could feel his smile all the way through the telephone line.

He was pleased, but told me, "This sounds good but there is still lots to do."

In Neil's last days I sensed the bottom line. He was satisfied with his life. He felt he had accomplished most things important. He could now permit himself to coast just a little and reflect on what had been and what must be.

In one of Neil's last visits to his Florida launch site, Herb Gold, NBC's associate producer for our space coverage team commented, "I had never seen him so relaxed. All times before, Neil had been uncomfortable, on guard.

"We sat down with you guys," Herb continued, "and as we enjoyed our libations, Neil asked my wife Sharon if she believed there were Martians among us."

Sharon smiled, but before she could answer, Neil told her, "There are Martians on Earth. They are our children in pre-kindergarten. They'll be the first Martians."

It was a gathering pretty much of the old NBC space coverage team including our boss Jim Kitchell, who would sadly write me when Neil passed, "At least we had those last moments with him."

It was indeed a good evening. When I and my wife Jo walked Neil to his accommodations at Cocoa Beach's Holiday Inn, the hotel we had all frequented in the Apollo days, Neil—not usually a touchy-feely person—hugged my wife, and then we put our arms around each other for a brotherly embrace.

To me it was an important road marker in the sunset of our lives. We both felt as if we'd been there, done that.

―――――――

A few days later Neil went off to spend a few weeks traveling the world with the last man on the moon, Gene Cernan, and his backup, the commander of *Apollo 13*, Jim Lovell.

They spent some time with our troops, sold space exploration to the movers and shakers, and served as America's ambassadors from space. All their travels were uneventful save one.

Their trip was almost over when Gene and Neil boarded an airliner from London to New York and Jim flew to his office in Chicago. Gene and Neil were flying a Boeing "triple-seven" scheduled to land at 6:00 P.M.—dinnertime in New York City. Being the dog days of summer it was also the peak time for thunderstorms, and the weather did not disappoint.

"There was this mother of all thunderstorms," Gene Cernan told me. "Lighting and turbulence we'd never seen before. It was big-time bad," he emphasized, "and the pilots, God bless them, tried to land in New York twice. It was no go.

"They waved us off to Boston, and once there put us on a bus back to New York.

"It was about one o'clock in the morning when the bus finally stopped and we all unloaded for a bite at a roadside diner. Neil and I were trying to make the best of it—two old astronauts sitting at the counter drinking our coffee and eating our Danish. Neil and I both sensed that the kid serving us didn't have a clue he was serving the first and last men on the moon."

Neil and Gene for a long time had not been recognized by the general public. Neil once told me he was only recognized at an event where he was expected. The bottom line was that neither cared. Public adulation was not a big thing for Neil and Gene. Their major concern was America's future in space—how we would get to the point where humans could travel safely to Mars.

Neil was convinced that once we knew how to live in space and ensure our personal survival, we would have taught ourselves how to reach Mars and more distant places in one spaceship—or a flotilla.

The question was, who should go? Will it be a global effort, or again a joint mission flown by two or three countries?

Will it be a return or a one-way trip? Will they follow the traditions of the wagon trains west in the 1800s and establish a colony, to be joined later by others, or will they return to Earth—to family and friends?

Whichever, it will be an exciting time, and Neil would have certainly liked to have been around for the twenty-first-century's greatest adventure.

But as stated, Neil had no regrets.

There really was only one annoying item he could never be done with—stamping out an outrageously fictitious story circulating in cyberspace for years.

The story, forwarded endlessly via e-mail:

As Neil Armstrong reentered the lunar module Eagle from his walk on the moon, he made the enigmatic remark—"Good luck, Mr. Gorsky."

Many people at NASA thought it was a casual remark concerning some rival Soviet cosmonaut. However, upon checking, there was no Gorsky in either the Russian or American space programs.

Over the years many people questioned Armstrong as to what the "good luck, Mr. Gorsky" statement meant—Armstrong just smiled.

On July 5, 1995, in Tampa Bay, Florida, while answering questions following a speech, a reporter brought up the 26-year-year-old question to Armstrong. This time Neil finally responded. Mr. and Mrs. Gorsky had died, so Neil felt he could now answer the question.

In 1938, when he was a kid in a small Midwestern town, he was playing baseball with a friend in the backyard. His friend hit the ball, which landed in his neighbor's yard by their bedroom window. His neighbors were Mr. and Mrs. Gorsky.

As he leaned down to pick up the ball, young Neil Armstrong heard Mrs. Gorsky shouting at Mr. Gorsky: "Sex! You want sex?! You'll get sex when the kid next door walks on the moon!"

The sad part is that millions today believe the ridiculous story to be true.

On June 7, 2012, little more than two-and-a-half months before Neil died, I wrote him:

Morning Neil:

I received this unbelievable yarn I've tried to knock down many times from my grandson Bryce, the football kicker you met at my 50th dinner.

I'm sure you have been pestered to death with many differing accounts of this tale.

For my kicker from your son the kicker, any comment?

Jay.

Neil replied:

Jay,

I first heard the story sometime in the 80s as told by comedian Buddy Hackett at a charity function.

As I'm sure you know the transcriptions of all the actual conversations are available on the Web at the Lunar Surface Journal.

I think there must be a secret club where they give Oscar-like awards for the most outrageous Internet scams in different categories: jokes, photographs, quotes, etc.

And there is a great deal of competition!

Best,
Neil

Rest in peace my good man. We'll be along directly.

INDEX